A PLACE FOR INQUIRY, A PLACE FOR WONDER

A Place for Inquiry,
A Place for Wonder

The Andrews Forest

WILLIAM G. ROBBINS

Oregon State University Press   Corvallis

Library of Congress Cataloging-in-Publication Data

Names: Robbins, William G., 1935- author.
Title: A place for inquiry, a place for wonder : the Andrews Forest / William G.
   Robbins.
Description: Corvallis, OR : Oregon State University Press, 2020. | Includes biblio-
   graphical references and index.
Identifiers: LCCN 2020027030 (print) | LCCN 2020027031 (ebook) | ISBN
   9780870710193 (trade paperback) | ISBN 9780870710216 (ebook)
Subjects: LCSH: Experimental forests—Oregon—H.J. Andrews Experimental Forest.
   | Environmental protection—Oregon—H.J. Andrews Experimental Forest. | H.J.
   Andrews Experimental Forest (Or.) | Forest ecology—Oregon—H.J. Andrews
   Experimental Forest. | Forest management—Social aspects—Oregon—H.J.
   Andrews Experimental Forest.
Classification: LCC SD358.8.O7 R63 2020 (print) | LCC SD358.8.O7 (ebook) |
   DDC 634.9/209795—dc23
LC record available at https://lccn.loc.gov/2020027030
LC ebook record available at https://lccn.loc.gov/2020027031

♾ This paper meets the requirements of ANSI/NISO Z39.48-1992
(Permanence of Paper).

Oregon State University
OSU Press

Oregon State University Press
121 The Valley Library
Corvallis OR 97331-4501
541-737-3166 • fax 541-737-3170
www.osupress.oregonstate.edu

Contents

This book is dedicated to the scientists, photographers, poets, essayists, and visitors who have appreciated the Andrews Forest as a special place.

Acknowledgments

Researching and writing the history of the H. J. Andrews Experimental Forest was a new enterprise for one accustomed to spending months in archives from Washington, DC, to the West Coast. Fresh from three years in Oregon State University's Special Collections and Archives researching and writing *The People's School: A History of Oregon State University*, I launched into the Andrews project with nary an archivist in sight. Except for a few seminal documents gleaned from the experimental forest's founding (the Andrews archival material was in the process of being inventoried), most of the scientific documents I sought were digitized listings in the massive Andrews Bibliography (2,806 documents and counting). Equally important to the research were published scientific materials available through Google Scholar. Still other articles were available online via the OSU Library. Incredible! No photocopying, no scanning, no note-taking. I did, however, purchase a new printer.

Many people who have spent their professional lives with the Andrews Experimental Forest have provided valuable support for the research and writing of this book. First and foremost is Fred Swanson, a wonderful and helpful friend, whose spiritual world is centered on the 15,800-acre forest. Fred guided a novitiate through the dense thicket of ecosystem science, critiquing my sometimes stumbling prose and, most important, offering cheerful support for my efforts. Fred also has a sharp eye for redundancies, word repetitions, and missing quotation and parentheses marks. Additional and important dimensions to Fred's association with the Andrews are the multitude of scientific articles that bear his name and his efforts to establish a place for the Andrews records with OSU's Special Collections and Archives. Fred, you are the best!

Among others who have guided me through the labyrinth of the Andrews story is Lina DiGregorio, friend, cul-de-sac neighbor, photographer, and coordinator of the Andrews Long-Term Ecological Research program. Lina's expertise is evident in many of the photos that appear in this book—and in selecting others from the hundreds of photos in the Andrews Image Library.

Sam Schmieding provided valuable service inventorying and organizing the extensive Andrews records. Forest Director Mark Schulze, who lives at the Andrews headquarters compound near Blue River, was always generous in responding to my occasional pesky questions over the last two years. Mark generously shared his insights about the forest when OSU president emeritus John Byrne and I visited the headquarters in September 2019. Michael Nelson, principal investigator for the National Science Foundation's Long-Term Ecological Research (LTER) grants since 2012, provided valuable observations about the present and future of the Andrews, especially its need to be on the cutting edge of ecosystem science in its grant applications.

Others who reflected on their experiences at the Andrews included Andrew Gray, research ecologist with the US Forest Service's Pacific Northwest Research Station, who offered insights about problems with adaptive management proposals under the Northwest Forest Plan. Richard "Dick" Waring, who, with Jerry Franklin, guided research programs under the early Long-Term Ecological Research programs, was helpful in the early stages of the study. Dick shared valuable insights about scientific research, establishing data record-keeping, and steered a non-scientist to the importance of Google Scholar. Mark Harmon, who followed Fred Swanson as principal investigator of the LTER grants in 2002, emphasized the importance of the Andrews as *place*, a special environment where eclectic groups of scientists engaged in creative inquiries about the functions of ecosystems. Barbara Bond, the only woman to serve as principal investigator for the Andrews, provided interesting commentary—the need for scientists to continually push the boundaries of creativity and inventiveness in their research programs.

Tom Spies, now emeritus research scientist with the Pacific Northwest Research Station, deserves special attention for guiding me through the murky world of the Northwest Forest Plan and its many-faceted programs since its implementation in 1994. Tom was helpful in clarifying the land allocations under the plan, and the acreages involved, and confirmed that I had the correct acreage figures for each section. Just before retiring from the Forest Service, Tom was the principal author of a major Forest Service study of the Northwest Forest Plan, *Synthesis of Science to Inform Land Management within the National Forest Plan Area* (2018).

En route to reading documents and writing, I spent a morning hour or so in coffee shops, two of them closing in early 2019. In finishing the writing, revisions, and other matters, I enjoyed the friendly staff at Susan's Garden and Coffee Shop.

A Note on Sources

Because of the uncertain repository for H. J. Andrews Experimental Forest Records at the time this book went to press, all copied documents are deposited in the William G. Robbins Papers, Special Collections and Archives, Oregon State University (hereafter, SCARC). All citations for oral history interviews are also located in the Robbins Papers.

Introduction

The human race is challenged more than ever before to demonstrate our mastery, not over nature but of ourselves.

—Rachel Carson[1]

In the spring of 1964—and still new to Oregon—I drove up the McKenzie River highway from Eugene to the town of Blue River, turned left along the stream of that name into the Willamette National Forest. After a few miles, I crossed a bridge and passed a sign along a smaller stream that read "H. J. Andrews Experimental Forest." Armed with a fly rod and can of worms, I fished Lookout Creek, a beautiful mountain stream shrouded in old-growth Douglas-fir and thick with an understory of vine maple and sword fern. Patience and keeping a low profile before the crystal-clear pools of water rewarded me with a nice catch of cutthroat trout. Several decades later and with a growing interest in environmental history, I became aware of the significance of "the Andrews," as it is known to faculty at my home institution, Oregon State University. The Forest Service's Pacific Northwest Research Station has offices on the OSU campus, and many of its scientists serve as adjunct faculty members.[2]

In the process of writing my first book, *Lumberjacks and Legislators* (1982), I learned that forestry research in the United States originated with Bernhard Fernow, a graduate of the Prussian Forest Academy in Germany, who served as chief of the Division of Forestry, from 1886 to 1898. Fernow published several investigations of forestry during his twelve years as chief and could be described as the grandfather of today's national forest system. He encouraged the federal government to withdraw its forest lands from public entry (prohibiting homesteads), so that they could "form a nucleus and an example for American forestry." Fernow's proposal moved toward reality when Edward Bowers, a lawyer in the General Land Office, drafted a legislative bill that became the Land Revision Act in 1891 (also known as the Forest Reserve Act). The measure granted the president authority to withdraw forest land from public entry to establish forest reserves, marking a dramatic shift in American land policy.

President Benjamin Harrison created six forest reserves in 1891–1892, and the following year, the new president, Grover Cleveland, withdrew the Cascade Range Reserve, from which the Willamette National Forest was created.[3]

Gifford Pinchot, who succeeded Fernow as chief of the Division of Forestry (1898–1905) and went on to become chief of the Forest Service (1905–1910), advanced forest research when he established the Section of Special Investigations in 1898 and directed his staff to carry out field surveys of the distribution and regeneration of commercially valuable trees.[4] Pinchot expanded the forestry bureaucracy, creating the Section of Silvics in 1903 "to contribute to ordered and scientific knowledge of our forests." There was no overall coordination of research until Raphael Zon, a brilliant Russian immigrant, joined the Forest Service in 1901. Zon headed the new Section of Silvics and then established the nation's first forest experiment station in Arizona's Coconino National Forest in 1908. Henry Graves, Pinchot's successor, established the Branch of Research in 1915 and appointed Zon to head the unit. Through the influence of Zon and others, Forest Service research became an integral component of the agency's responsibilities, providing information on economic conditions in the lumber industry, conducting experiments with seed germination in nurseries, and studying reforestation practices.[5]

Zon was indirectly involved in establishing the first forest experiment station in the Pacific Northwest. When the Forest Service sought a location for a research station in the Douglas-fir region, the agency tabbed Thornton Munger to head the new Wind River Experiment Station in 1912. A young graduate forester with several months' experience in Zon's Washington, DC, office, Munger had been working in the region since 1908. Although the Wind River Experiment Station carried out wide-ranging research, the Wind River Experimental Forest was not formally established until 1932. Scientific investigations in the Wind River Valley, however, began with Munger's arrival in 1908. At approximately 10,000 acres, the Wind River Experimental Forest has been home to the oldest forestry research in the Pacific Northwest. To provide oversight for research in the agency's Region 6 (Oregon and Washington), the Forest Service established the Pacific Northwest Forest and Range Experiment Station in Portland in 1925. By its half-century mark in 1975, the station's field offices and laboratories were scattered across nine distant sites, its central offices in Portland employing three hundred people.[6]

To flesh out the distribution of experimental forests across dominant forest types in the Pacific Northwest, the Forest Service established the Pringle Falls Experimental Forest (ponderosa pine) in 1931, Cascade Head

Experimental Forest (Sitka spruce, western hemlock) in 1934, Port Orford Cedar Experimental Forest in 1934 (discontinued in 1958), South Umpqua Experimental Forest in 1951 (mixed evergreen forest type); Blue Mountain Experimental Forest (ponderosa and inland pine, merged with the Starkey in late 1950s), and the Starkey Experimental Forest and Range (mixed forest and range) in 1941. The Pacific Northwest Forest and Range Research Station served as the nerve center for the experimental forests. Of the experimental forests in Region 6, the H. J. Andrews Experimental Forest, originally established as the Blue River Experimental Forest in 1948, emerged as the preeminent leader for its wide-ranging, complex, and sophisticated research in ecosystem science, winning national and international acclaim.[7]

The Andrews Experimental Forest resurfaced in my life when I attended a seminar in the campus offices of the Pacific Northwest Research Station in the mid-1980s. On that occasion, scientists Chris Maser and James Trappe, and a colleague, presented research findings explaining the importance of downed trees, small animals, and microbial life to the health of a forest. They described problems with European forestry wherein the forest floor had been swept clear of woody debris for decades, depriving trees of nitrogen and minerals necessary to sustain plant life. The event was an eye-opener for one who still believed in the efficacy of Forest Service plantation forestry, cutting old-growth forests and replanting them to fast-growing young stands. Although not explicitly scientific, a few of my articles and two early books attracted the attention of Fred Swanson and Jim Sedell, both early participants in Andrews research.

My next association with the Andrews took place in 2015 while writing a brief commentary about the experimental forest in *The People's School: A History of Oregon State University*. I cited the research carried out at the Andrews Experimental Forest as an example of the influence of the environmental age in framing new approaches to forest science, referring to its investigations "as an evolving collaborative laboratory for multiagency research."[8] As the writing and revisions to that manuscript were winding down, I began meeting with Fred Swanson (when he was having the oil changed in his vehicle at the local Honda dealership) about the possibility of bringing public attention to the Andrews, a place virtually unknown to Oregonians (and most environmental historians). While the book manuscript was out for review, I immersed myself in secondary literature about the Andrews. Through those readings, I learned that the forest's website made available to the public a large trove of impressive digitized documents.

Fred Swanson, a geomorphologist with a PhD from the University of Oregon, spent most of his professional career as a research geologist with the Pacific Northwest Research Station in Corvallis, where his eclectic energies focused on ecological responses to disturbances, land-use legacies, and the influence of science on natural resource management and policy. Swanson is a principal founder and custodian of the Long-Term Ecological Reflections Program in the Arts and Humanities and their relation to science.

By the time *The People's School* appeared in print, I was preparing an article-length manuscript about the Andrews for the *Oregon Historical Quarterly*.[9] At that point, I was making progress in getting a handle on terminology in the multitude of research articles related to scientific investigations on the experimental forest. Although its reputation as a place for breakthrough research was long in gestation, its emergence as a center for ecosystems inquiry was clearly related to the advent of the environmental age, the publication of Rachel Carson's *Silent Spring* (1962), congressional passage of the National Environmental Policy Act (NEPA) in 1969, the Endangered Species Act in 1973, and the National Forest Management Act of 1976. NEPA mandated the federal government to "create and maintain conditions under which man and nature can exist in productive harmony."[10]

The Andrews Forest is one of four experimental forests in Oregon. As part of the Willamette National Forest, the Andrews is a thoroughly western Oregon place of steep mountain slopes, narrow valleys, and cascading streams that empty into Lookout Creek, the main tributary to Blue River. The Andrews Experimental Forest's reputation was long in the making. Its history includes participation in the International Biological Program (IBP) between 1968 and 1974 and the National Science Foundation designating it as one of its Long-Term Ecological Research (LTER) sites in 1980. Beginning in 1993, the forest was directly linked to the Northwest Forest Plan, a huge effort involving national forests in western Washington and Oregon and northern California. Andrews scientists have been involved with contentious issues related to protecting the northern spotted owl and offering modest timber sales to dependent communities.[11]

The H. J. Andrews Experimental Forest (15,800 acres) and the Lookout Creek drainage that defines its boundaries once encompassed great stands of old-growth conifers ranging from four hundred to five hundred years old, a place reminiscent of forests in the western Cascades after the Second World War. From its founding in 1948, the experimental forest has been the setting for ever-expanding fields of research. Beginning with postwar studies focused on converting the huge volume of old-growth timber to fast-growing young stands, Andrews research evolved over the years to long-term ecosystems investigations that continue to the present day. Over time, those inquiries profoundly reshaped Forest Service management policies and sharpened our knowledge

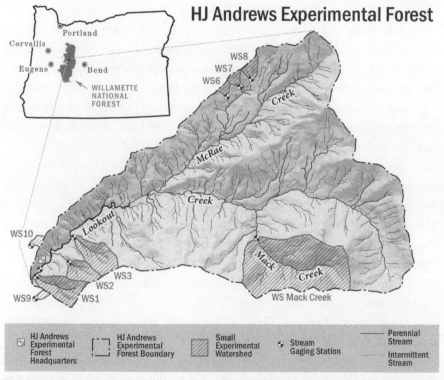

The H. J. Andrews Experimental Forest is located in the Willamette National Forest and is one of four Forest Service experimental forests in Oregon. This map indicates the forest's boundaries (the Lookout Creek drainage), principal streams, and small experimental watersheds and gaging stations. Map by Jesse Netter, courtesy of the *Oregon Historical Quarterly*.

about healthy forest environments—a dramatic shift in federal timber practices from an industrial paradigm involving intensive forest management to policies emphasizing biodiversity and healthy ecosystems. The title of Jon Luoma's book, *The Hidden Forest* (1999), is a fitting description for the site of the work that has taken place on the Lookout Creek drainage. The absence of signage on highway 126 contributes to the anonymity of the place.[12]

The creation of the Blue River Experimental Forest occurred amid the skyrocketing demand for building material after the Second World War. The collapse of the construction industry during the Great Depression and civilian savings during the war unleashed an insatiable need for building supplies in the 1950s and beyond. The long-deferred home construction industry flourished, and sawmill communities in the timber-rich Pacific Northwest boomed. Because private timber harvests were unable to meet consumer needs, the escalating plea for timber pressured the Forest Service to increase sales. The agency, which had been operating largely in a stewardship mode for decades, mostly selling select cuttings to local communities, did an about-face, transitioning to clear-cutting and fully embracing market-induced production strategies. The Forest Service aggressively promoted converting old-growth forests into fast-growing young stands of trees, adopting intensive forestry practices based on the idea that advances in forest science would meet the demand for wood products.[13]

With its soaring production records, the Pacific Northwest timber industry was in liftoff, with Oregon reaching an all-time high of 9.1 billion board feet of lumber in 1955, 8.2 billion in 1965, and 7.3 billion in 1979. That unprecedented construction boom, which contrasted sharply with the morbid industrial economy during the Depression, persisted until the early 1980s when a severe recession in home building caused mill closures in Oregon and timber-dependent communities across the region. The establishment of the Blue River Experimental Forest at the onset of the surge in timber sales significantly influenced the direction of research for nearly three decades.[14] Although future research would address the functions of ecosystems, investigations through the 1950s and 1960s focused on "conversion," a mantra for the industrial paradigm: to efficiently harvest old-growth trees and establish new stands of timber.[15]

The initiative to set aside a significant body of forest within the Willamette National Forest in 1948 for research represented the insights of Horace J. Andrews, the forester in charge of Forest Service Region 6, which encompassed Oregon and Washington. Andrews wanted to better understand the effects of logging on water quality and fisheries habitat, and the influence of the Cascade Range on flooding in the Willamette Valley. And there was more to Andrews's worries, especially the need to protect streams and rivers in the face of the expected increase in timber harvests. A cataclysmic event, the Columbia River flood in the late spring of 1948, loosened federal purse strings to support watershed research. A strong supporter of forest research, Andrews was directly involved in selecting the Lookout Creek drainage near the community of Blue River as the location for the experimental forest. After his death in an automobile accident in Washington, DC, the experimental forest was renamed in his honor in 1953.[16]

Among the early investigations on the Andrews, none would have greater significance than watershed studies. Robert Cowlin of the Pacific Northwest Research Station indicated that the purpose of the experimental forest was to better understand the relationship between silvicultural practices and healthy watersheds. Those suggestions represented the efforts of Jerry Dunford, who drafted plans to study the effects of logging on three small watersheds near the mouth of Lookout Creek, focusing on stream fluctuations, water quality, snow accumulation, and moisture in the soil. The watershed investigations, designed to last "through one rotation" (harvest cycle), were critical to decades of research on the Andrews. Dunford ordered the installation of "three trapezoidal flume stream gauges with recorders," silt traps, and rain gauges to obtain an understanding of soil disturbances on watersheds.[17]

The H. J. Andrews Experimental Forest first gained national and international prominence when the National Science Foundation selected it to participate in the International Biological Program (IBP) in 1969.[18] Equally important for the future of the Andrews and other sites seeking IBP funding, several of them were on national forests. From their association with the International Biological Program, Andrews scientists moved seamlessly to the National Science Foundation's designation of the experimental forest as one of its first Long-Term Ecological Research sites. The association with NSF enhanced the shift toward collaborative and interdisciplinary investigations, a marked transition from the earlier bifurcated, isolated, individualistic biological research following the Second World War.[19]

Since the Forest Service established the experimental forest in 1948, scientific terms for research have changed. Like the national forest system itself, the words ecosystems and ecosystem studies are absent from Andrews documents well into the 1960s. Research agendas in the 1950s refer to a "proposed treatment schedule" and "treated watersheds," with findings to "insure as rapid regeneration of the timber on this area as is reasonably possible." From unapologetic, applied approaches to investigations, the terms and focus of research evolved to the point that scientists began referring to a forest as an ecosystem, a component of larger, region-wide ecosystems.[20]

The Andrews Forest and the Willamette National Forest bridged the transition from an exclusive focus on timber production to addressing biodiversity and related issues following federal District Judge William Dwyer's injunction in May 1991 blocking federal timber sales involving potential habitat for the northern spotted owl.[21] Dwyer's decision ultimately led to dramatic reductions in timber harvests on federal forests and affected research activities involving Andrews scientists who were collaborating with national forest managers. Judge Dwyer's verdict is a reminder that the world of courts and politics were never far from activities on the Andrews Experimental Forest. Liberal presidential administrations usually fostered a legislative environment more favorable to scientific research related to ecosystems management.

The Wilderness Act (1964) enabled Congress to set aside lands that would be off-limits to development; the National Environmental Policy Act (1969) required public input in land-management decisions; the Endangered Species Act (1973) mandated the protection of endangered wildlife and their habitats on federal land; and the National Forest Management Act (1976) required each national forest to create detailed plans (with public input) and opened the window to more restrictive environmental policies. The significance of those major legislative authorizations often waxed and waned with the nation's political mood. Activities involving scientific investigations on the nation's forests, wildlife refuges, national parks, and public waterways depended on the support of the executive branch and Congress.[22]

From the passage of the Wilderness Act in 1964 through President Jimmy Carter signing the Alaska National Interest Lands Conservation Act before he left office in 1980, Congress passed several environmental laws that would largely stand the test of time. Those legislative initiatives invited strong opposition, especially from Republicans, although loyalty was likely directed to home states and their major natural resources interests, rather than political parties. The influential Republican senator Mark Hatfield from Oregon supported many environmental laws but opposed all logging restrictions on public lands.[23]

During President Jimmy Carter's administration (1977–1981), market incentives began to play a greater role in regulatory agencies, reflecting a national trend toward more conservative politics. The coming of Ronald Reagan to the presidency accelerated that trend. With Reagan's inauguration in January 1981, twelve new Republicans joined the Senate, giving them control of the upper house. Although there was little difference between Reagan and Richard Nixon in their opposition to environmental legislation, Reagan's appointees to strategic federal offices were aggressive in championing states' rights and deregulation. He appointed John B. Crowell, a timber-industry lobbyist, as assistant secretary of agriculture to oversee the Forest Service. James Watt, who headed the ultraconservative Mountain States Legal Foundation, became secretary of the interior, where he set out to weaken environmental laws affecting public lands in the American West. When Watt and brewery magnate James Coors succeeded in promoting their fiery Colorado associate Anne Gorsuch to head the Environmental Protection Agency, Gorsuch and Watt worked in tandem to lessen the regulatory agencies under their charge.[24]

At the end of Reagan's two-term presidency, the administration's efforts had failed to destroy the regulatory powers of the EPA and other federal agencies charged with environmental oversight. Reagan officials, however, succeeded in weakening agency regulatory powers through budget processes, primarily through significant reductions in staff. By the end of his first term, Reagan's policies redirected environmental policy to state-level decision-makers. There were also frequent stories about administrators manipulating scientific data and screening the ideological positions of scientific advisory appointments. In the end, the Reagan "revolution" used presidential powers to reduce taxes, spending, and regulations. Under Anne Gorsuch, the EPA experienced enormous budget cuts, reducing the agency's payroll and its headquarters staff.[25]

The carryover from Judge William Dwyer's northern spotted owl decision led to President Bill Clinton's Forest Conference in Portland in April 1993, where scientists affiliated with the Andrews provided critical testimony. The outcome led directly to the adoption of the Northwest Forest Plan the following year. The Northwest Forest Plan, which is treated throughout this book, mandated the protection of designated old-growth stands and certain riparian areas.[26] For the next two decades and more, Andrews scientists were heavily involved in multifaceted, complex experimental programs under the plan. Timber industrialists and environmentalists opposed some of the plan's experimental strategies: industry because they drastically restricted national forest and Bureau of Land Management harvests, and the environmental community because they permitted limited cutting of old-growth forests for experimental purposes. These were trying years for scientists who proposed forest experiments that would yield wood fiber and nurture the development of old-growth characteristics in young timber stands. The Andrews Forest and the greater Blue River landscapes were central to those investigations.

Attention to threatened and endangered species and their habitats prompted major shifts in ecological science, away from focusing on single species to large ecosystems, from the plight of the owl to the larger physical environment necessary for the bird's survival. Through their growing interest in larger landscapes, Andrews scientists increasingly focused on ecosystems, a metaphor for an area of science that lacked specific definition. Despite the persisting fuzziness of the term, Forest Service research stations, the academic community, contributors to the *Journal of Forestry*, and scholars outside forestry profession advanced ideas about the significance of ecosystems at much larger spatial scales. The concept had fierce opponents as well, many of them in the industrial world, and also an academic provocateur, Alston Chase.[27] Those discordant voices did not deter the investigations of scientists, especially Jerry Franklin, who was in the forefront of scientists speaking out and publishing articles on the efficacy of ecosystem management.[28]

Franklin and many other scientists, who became central figures in the growing reputation of investigations on the Andrews Forest, are an important part of this story, because science is always dependent on the personalities involved. It is they who took interest in the plant and animal life on the Lookout Creek drainage, who opened inquiries into terrestrial/stream relations, and who discovered the nexus between the spotted owl and its habitat requirements. This book accentuates the significance of the relationships among personalities, the subjects they chose to study, and, for most of the scientists who plied the streams and slopes of the Andrews, their affection for the place. As Harvard Forest's David Foster remarked, "discussing science without this personal dimension ignores the larger and often intriguing story of how studies were undertaken and how lessons were actually learned."[29]

The Andrews Experimental Forest today is a remarkable and friendly place to visit for scientists and visitors alike. The headquarters compound itself has all the amenities of a small community, including problematic cellphone reception. Educational groups and special interest organizations, such

H. J. Andrews headquarters. Photo by Al Levno, January 1998.

as senior citizens, regularly make the long trip to take part in the delights of the forest. The headquarters location offers opportunities to observe unique and important research projects—a short trek to a stand of immense old-growth Douglas-fir, a tree with ropes where qualified individuals can climb to various heights to read temperature and moisture gauges. At another station is the headquarters central, benchmark meteorological equipment, where precipitation and air temperature, snow, wind speed, solar radiation, and precipitation data has been recorded for decades.

Visiting graduate students and scientists from around the world spend time on the Lookout Creek drainage pursuing investigations on meteorological data (precipitation and temperature differentials at various elevations and aspect in the forest), research into natural and human disturbances over long periods of time, carbon sequestration differentials between old-growth and younger forest stands, and biodiversity comparisons in older and younger forests. In recent years, researchers are paying greater attention to the effects of climate change—historic information on air temperature, precipitation (rain and snow), soil temperatures (in old and young forests)—some of the measurements dating to the 1950s. The Andrews website (andrewsforest. oregonstate.edu/about) describes the Lookout Creek drainage as "a center for forest and stream ecosystem research in the Pacific Northwest." It is that and much more—a place of more than one hundred bird species, numerous reptiles and amphibians, and mammals large and small, all of them on a landscape shrouded in seasonal fog, rain or snow (depending on elevation), old-growth trees still covering 40 percent of its 15,800 acres, and humanities scholars attempting to explain its wonders.

Chapter 1
Foundations

The highest service that forests of the Douglas-fir region can render is in support and stabilization of communities dependent on them. Included are, not only the people and investments in forest industries, but also those in farms, stores, banks, garages, schools, transportation, and various industries.

—H. J. Andrews[1]

The H. J. Andrews Experimental Forest is a thoroughly western Oregon place of narrow forested valleys, steep mountain slopes, and cascading streams. The Lookout Creek drainage that defines its boundaries and its great stands of old-growth conifers ranging from four hundred to five hundred years old are reminiscent of much of western Oregon after the Second World War. Tragedy struck Region 6 when well-liked head forester Horace J. Andrews was killed in an automobile accident while house-hunting in Washington, DC, where he was being transferred to an important Forest Service position in the agency's central offices. Fifty-nine years old at the time of his death, he was considered one of the leading Forest Service administrators in understanding Northwest forestry issues.[2]

Born in Michigan in 1892, Andrews earned bachelor's and master's degrees in economics from the University of Michigan in 1915 and 1916. Before joining the Pacific Northwest Forest and Range Experiment Station in Portland as senior forest economist in 1930, he worked seasonally on the Santiam (now the Willamette National Forest) in Oregon, the Arapahoe in Colorado, and the Lassen, Stanislaus, and Sierra National Forests in California. His résumé included stints as a surveyor for private lumber companies, a lengthy tenure with the Michigan Department of Conservation (eventually becoming state fire warden), and teaching appointments at Iowa State College and the University of Michigan. Assistant Forester Earle Clapp, who headed the Forest Service's Branch of Research in the Washington office, asked Andrews to assume responsibility for overseeing the big Forest Service survey of the nation's forest resources, to begin in 1930.[3]

In his November 1929 letter, Clapp urged Andrews to resign his appointment with Michigan's Department of Conservation and hurry to Portland for the new position. Before the month was out, Thornton Munger, director of the research station, confirmed Andrews's appointment as senior forest economist in charge of the forest survey. His supervision of the project would be important to the success of the enterprise. Working under Munger, Andrews oversaw the vast surveying operation covering both public and private forests in Oregon and Washington. From all indications he was well-liked and important to the success of the undertaking. When Chief Forester Ferdinand Silcox later wanted to transfer Andrews to a more prominent position in Region 6 with a higher salary, Clapp protested the move. He wrote Munger about Andrews's importance to the survey and that he "could not be spared without serious detriment to the whole Forest Survey project." Clapp added that the Washington office was undergoing "sweeping changes" that would involve salary readjustments to obtain another $5,600 salary allocation for Andrews. Because Andrews remained with the Northwest forest survey until 1939, we can assume that Clapp's remonstrances were successful.[4]

H. J. Andrews was a forester of substance and similar to other progressives who followed Gifford Pinchot into the profession. That Earle Clapp took an interest in promoting Andrews to head the forest survey in Region 6 is no surprise, given Clapp's own background in encouraging socially responsible research and forest stewardship in the agency. Like Andrews, Clapp held forestry degrees from the University of Michigan. During his first tour in Region 6, Andrews participated in meetings of the Oregon Planning Council, a state affiliate of President Franklin D. Roosevelt's National Planning Board, established in 1933. In October 1934, Andrews delivered a presentation to the Oregon council's "Land Planning and Use Committees," supporting land classification to achieve greater efficiencies in land use. Addressing agricultural and forestry issues, he urged zoning policies to achieve efficiency in land occupation and use. Focusing on tax-forfeited lands under county ownership, Andrews preferred state control, wherein tenure would be more stable. In cases where private ownership was failing, he urged the state planning council to seek ways to assist owners.[5]

When he addressed the Pacific Northwest Regional Planning Conference two months later, Andrews emphasized the benefits of integrating land uses, "the sum and substance of land use planning." He believed planners were successful

Horace J. Andrews was instrumental in establishing the Blue River Experimental Forest in 1948 when he was in charge of Forest Service Region 6. When Andrews was killed in an automobile accident in Washington, DC, the forest was renamed in his honor.

when they had integrated all uses into a "harmonious whole beneficial to both the land itself and the users of the land." Rotating and diversifying crops to keep land productive made sense in the long run, but the problem rested with humans who were "interested only in the immediate present." Andrews told the audience that planners tended to think only about land's proper uses and ignored the need to caution landowners about seeking only profits from land.[6]

Andrews left the forest survey in the summer of 1938 to accept a position at the University of Michigan as Charles Lathrop Pack Professor of Wild Land Utilization. This was a research appointment, with other responsibilities limited to directing "problem work by individual students" and participating in a land-use seminar. In a December 1938 letter to forestry dean Samuel Trask Dana, Andrews summarized his work surveying land use, and discussed the effectiveness of public relations work with citizens and why legislators made decisions on conservation measures. All seemed to be going well for Andrews until his blunt two-paragraph letter to Dana on July 17, 1939, informing the dean that he was severing his ties with the university and returning to Oregon "due largely to the fact that my family prefers living on the West Coast to living in Ann Arbor."[7]

He left Ann Arbor for Forest Service Region 6 in Portland, where he would be in charge of state and private forestry, one of the agency's three missions.

His duties would include overseeing federal and private cooperation in sustained-yield programs and state and private cooperation in fire protection under the Clarke-McNary Act of 1924. When Region 6 head Lyle Watts was appointed chief of the Forest Service in 1943, the agency elevated Andrews to regional forester for the Pacific Northwest Region (Region 6), a position he would hold with distinction until his death in 1951. Although Watts's signature is on the appointment letter, one suspects that Earle Clapp, who had been acting chief since 1939, was involved in the decision.[8]

The timing of Andrews's return to the Pacific Northwest and Watts replacing acting Forest Service chief Clapp in the same year (1943) is a bit ironic. Because of Clapp's progressive views, President Roosevelt was unable to have him formally appointed chief when Ferdinand Silcox died in 1939. Opposition from timber industry leaders, especially those from the Pacific Northwest, the most productive lumber-producing region in the country, made Clapp's appointment problematic. Industry leaders were incensed with the release in 1933 of the US Senate's *A National Plan for Forestry* (under Clapp's authorship) and its undertones of regulating forestry practices on private land. When Clapp retired in 1945, Andrews praised him for being "a strong driving force" in promoting research in the agency where, he said, "you have probably contributed more . . . than anyone else."[9]

With the assistance of several Forest Service personnel, especially Philip Briegleb, Region 6 chief Andrews lobbied successfully to establish the Blue River Experimental Forest in 1948. Andrews, who reasoned correctly that the postwar building boom would accelerate timber cutting on national forests, was concerned that increased harvests would contribute to flooding and be detrimental to downstream water quality and fishing. When acting Forest Service chief Richard McArdle signed the agreement creating the Blue River Experimental Forest, he emphasized its purpose—"the conversion of these over-mature forests to managed young-growth stands in the most orderly manner with the least delay." McArdle's statement underscored the prevailing forest science of his generation: an exclusive focus on the association between on-the-ground investigations and forest productivity. The Lookout Creek watershed, the principal tributary of Blue River, would serve as a laboratory for Forest Service research to assure that the process of conversion would be efficient.[10]

Willamette National Forest supervisor J. R. Bruckart, Pacific Northwest Forest and Range Experiment Station director J. Alfred Hall, and Region 6 forester Andrews signed the agreement establishing the Blue River Experimental Forest. Although the forest would operate under the Northwest Experiment Station, logging on the Lookout drainage was to fit "the requirement for timber removal planned by the Willamette National Forest." The Northwest station director was to cooperate with the Willamette supervisor in determining the volume and methods of harvesting. The Blue River Experimental Forest would remain an integral segment of the McKenzie "working circle," a spatial strategy for determining the annual cut for a given area. The experiment station would also cooperate with the Army Corps of Engineers' Blue River Snow Laboratory adjacent to the experimental forest, where logging was prohibited until 1957. In drafting the plan for the experimental forest, Northwest station director Hall adjusted Blue River's research program to assist the timber industry in more efficiently logging old-growth timber.[11]

Other stipulations in the forest's initial agreement emphasized the importance of logging: "In accepting access road money for opening up the Blue River watershed, it is necessary that the full cut for the next 15 years come from the Lookout Creek drainage." If the experimental forest failed to meet "the size [of cutting] contemplated by the National forest, the Supervisor may then make regular timber sales to meet the planned cut on the watershed." Although no headquarters buildings were planned, the experimental forest would have a trailer house at the Blue River Ranger District buildings in McKenzie Bridge. In light of future path-breaking research on streams and animal life on the Lookout drainage, the agreement did not list investigations involving fish and wildlife.[12]

The economic and political environment that drove early research on the experimental forest reflected the Forest Service's need to provide logs primarily for the big lumber mills in Eugene and Springfield. "The Andrews," writes Jon Luoma, "was to be a key industrial laboratory." As the twentieth century advanced, the function and purpose of the forest would shift dramatically away from its founding objective—conversion—to issues of much greater scientific complexity. Forest Service documents of the time, however, described the Lookout drainage as sufficiently large to provide "answers needed for managing entire watersheds . . . [and] to test logging methods and techniques on commercial-sized operations."[13]

In cooperation with the Army Corps of Engineers, which was in the midst of building flood control dams on the Willamette River system, the experimental forest was commissioned to investigate forest influences on "streamflow, run-off, snow melt, and other hydrology." A cataclysmic event, the great

Columbia River flood in the late spring of 1948, provided a powerful incentive for the federal government to invest in watershed studies. Although most of the destruction was far removed from the McKenzie River drainage, the flooding Columbia loosened federal purse strings for funds channeled through the Pacific Northwest Research Station.[14] Within a brief period, those funds set in motion streamflow investigations on the Andrews Experimental Forest that would continue to the present day.

The postwar ambitions of the Willamette National Forest placed great pressure on Roy Silen, the first forester in charge of the Andrews, to meet harvest objectives. The experimental forest's management plan emphasized "large-scale experimental cuttings." Projected harvests would be twenty million board feet for the first decade to fit the Willamette's harvest plans for the Blue River Ranger District. Issued in March 1950, the "Working Plan" for the first experimental forest sales listed six clear-cut units, with an anticipated volume of 23,500,000 board feet. Road construction would be designed to fit topographic conditions and timber stand characteristics, with a careful eye to cost and efficiency.[15]

Critically important to future scientific investigations on the Andrews, the working plan included watershed studies, "one of the primary purposes of the Blue River Experimental Forest." Stream gauges had already been installed on Blue River and Lookout Creek, and because road-building contributed to most of the sediment reaching streams, road location was critical to limiting erosion. In an observation that continues to bedevil stream ecologists, the plan stated, "It takes very little grading with a bulldozer in a stream to muddy the water for miles." It was imperative, therefore, to properly lay out timber sales and road construction.[16]

Roy Silen was operating in a Forest Service helter-skelter environment of increasing harvests, inventorying standing timber, and developing predictive models for second-growth timber. Silen, who grew up in Coos Bay on the southern Oregon coast, attended Oregon State College, where he studied forest management and logging engineering. After a stint in the army during the Second World War, he hired on with the Pacific Northwest Forest and Range Experiment Station and worked under well-known Thornton Munger just before the latter retired. Silen lived through the Forest Service's transition from a stewardship/protection agency to full-blown timber production following the war. He associated with others at the Northwest Research Station who insisted that federal foresters, not loggers, should develop timber sales on

Roy Silen was the first
forester in charge of the Blue
River Experimental Forest.
Silen, with the Pacific
Northwest Research Station,
laid out the first roads and
timber sales in the Lookout
Creek drainage. In 1954 the
station transferred him to its
offices on the OSU campus,
where he headed the new
forest genetics program.

national forests and establish the location of roads and landings. With Robert
Ruth, Silen published *Getting More Forestry into the Logging Plan*, a best seller
at the station for several years.[17]

When Silen arrived at the new Blue River Experimental Forest in 1948,
access to the Lookout drainage extended only a mile or two beyond the
town of Blue River: "You had to walk before you got to the boundary of the
experimental forest," he told an interviewer. During his first summer and into
the fall season, Silen lived in a twelve-foot trailer parked in the old Civilian
Conservation Corps camp at the location of today's McKenzie River Ranger
Station. From there, Silen made weeklong backpacking trips through the
Lookout drainage, sometimes with his assistant, Hank Gratkowski, staking
out sales units in "chopped up country" with "very steep slopes." Silen was
proud of his work designing roads and laying out logging sales, finding "a way
to reduce the sedimentation in streams."[18]

Silen considered his most significant responsibility to be examining
"streamflow from an uncut drainage," that is, from an area never before harvested. His preliminary observations suggested that road construction was the
major problem because of the danger roads posed for increasing landslides.
His reports also reveal the beginnings of interdisciplinary research on the

Andrews, practices that would become the hallmark of future investigations. In addition to Army Corps of Engineers studies of streamflow and snowpack, the Oregon Cooperative Wildlife Research Unit spent three summers on the forest examining the effects of logging on fish and wildlife. An updated memorandum of understanding (MOU) with the Willamette National Forest in 1953 reduced the annual cut on the Andrews from twenty to seven million board feet. Important for the future, according to Silen, was "the complete exclusion of the small watershed study areas from consideration in the cutting program," an early acknowledgement of their potential for research. The agreement referred to three watersheds near the mouth of Lookout Creek.[19]

H. J. Andrews's death in an automobile accident in Washington, DC, in March 1951 occurred when he had been promoted to assistant chief in charge of national forest administration in the nation's capital. His passing at age fifty-nine sent shockwaves through the professional forestry world and beyond. Although the obituaries were legion, one refrain runs through them all: a man who was kind to a fault, well regarded in the industry and the Forest Service as well. The leading industrial journal, the *Timberman*, praised Andrews for being "able to translate to industry the perplexing matters of policy that must govern in the administration of any public property." The *Timberman* quoted the popular writer Stewart Holbrook's tribute on the editorial page of the *Oregonian*: "As a person Horace Andrews was a genial soul, a staunch friend in rain or shine, and a man who knew there were unqualified answers to almost no questions." Associate Justice William O. Douglas of the US Supreme Court penned a note to Mrs. Andrews about his admiration for her husband and sent her his "deep and abiding affection as well." He was, Douglas said, "as fine a public servant as the nation has known."[20]

Sixteen months later, Forest Service royalty, friends, and family gathered at a picnic area just inside the experimental forest on July 26, 1953, to dedicate the place in honor of H. J. Andrews. Andrews's successor as Region 6 forester, J. Herbert Stone, presided at the event. Invited speakers included Lyle Watts, the retired chief of the Forest Service; Robert Cowlin, director of the Pacific Northwest Research Station; and Oregon's state forester, George Spaur. A picnic lunch was followed by the formal dedication, and then guests were invited to tour the experimental forest. At the time of the dedication, the forest had sixteen miles of road, and old-growth timber had been harvested on eighteen plots. In the same month, the *Oregon Journal* reported that researchers were investigating reforestation problems on cutover lands, developing management practices to provide "pure water" for downstream towns and industries, and reducing losses of trees from wind, insect, and diseases.[21]

Many of the progressive changes that trickled down to the Andrews reflected Jerry Dunford's return to the Pacific Northwest Research Station from the Rocky Mountain Region in 1952 to oversee watershed questions. With the assistance of W. E. Bullard, Dunford drafted the "Work Plan for Forest Influences Studies," emphasizing the importance of watershed investigations. Although the plan appeared in the memorandum of understanding of 1953, no immediate initiatives were taken. Dunford, who proposed to study soil erosion and sediment in streams, ordered the installation of sophisticated instruments, "three trapezoidal flume stream gauges with recorders," silt traps, and rain gauges on three small watersheds on lower Lookout Creek. His proposal was practical: to understand how soil disturbances affected watersheds, with the proposition that carefully planned roads and landings would lessen the loss of soil. His broader objective was to develop strategies to lessen the effects of timber harvests on water quality and streamflow. Dunford believed the data collected would provide a better understanding of how to protect streams from harvest operations. Studies already indicated that carefully mapped landings and skid trails would lessen soil disturbances.[22]

For reasons that are not clear, the Pacific Northwest Research Station reassigned Roy Silen to head the forest genetics program at Oregon State College in 1954. His position at the station remained vacant until Jack Rothacher assumed the post in 1957. From all indications, Silen's leaving was not pleasant. He clashed with Forest Service engineers who did not like his road-building layouts. Engineers would tell him that his designs did not meet their road standards, an accusation that prompted Silen to charge, in an interview, that they "were forcing us to do dumb things." Blue River Ranger District officials were obviously pushing strategies—especially for building less-expensive roads—that differed from Silen's layout. When interviewer Max Geier asked Silen decades later whether he had had any involvement with the Andrews after his transfer, Silen responded that he was "just chopped off" and had no contact with the staff on Lookout Creek. When he visited the Andrews in the mid-1990s, he commented that "it looked pretty good."[23]

Substantial changes came to the Blue River Ranger District in 1957, when the Forest Service built new headquarters in the town of Blue River. To facilitate the relocation, the Willamette National Forest and the Pacific Northwest

Research Station signed a new MOU detailing accommodations and other considerations for experimental forest personnel. The Andrews staff would "occupy the smaller of the rooms in the basement of the new office at Blue River." The research station, still without a finished interior, was responsible for finishing the room and paying for heating, electricity, and telephone charges. Although the office was in an unfinished basement, the building was a considerable upgrade from the original house trailer.[24]

External events disrupted the experimental forest's plans when the Willamette National Forest postponed an Andrews timber sale, because the Willamette was fast-tracking harvests along the South Fork of the McKenzie River to clear the way for Cougar Reservoir. When dam construction blocked upstream access a few months later, the national forest supervisor "proposed a greatly accelerated sales program on the Andrews to help maintain the allowable cut in the McKenzie working circle." A revised MOU, in 1959, again emphasized the limited range of options for the Andrews—cutting could "be accelerated, reduced, or stopped when necessary" to meet the objectives of the Willamette National Forest.[25]

After delaying for three years, the Pacific Northwest Research Station—at Jerry Dunford's suggestion—hired Jack Rothacher, a former district ranger on the Umpqua National Forest, to fill Roy Silen's position. Rothacher, who remained with the Andrews until 1974, was a capable administrator, fostering cordial relations among scientists, the Blue River Ranger District, and Willamette National Forest officials. His first major report in 1958 repeated conventional research objectives—"to learn and demonstrate improved methods of multiple-use management for old-growth Douglas-fir forests on mountain watersheds," and to develop "vigorous young-growth stands." The three small watershed studies, Rothacher reported, had collected five years of streamflow and precipitation data that were important, because forested watersheds needed "to produce a maximum of timber, without detriment to the supply of useable water." Like his peers in the watershed investigations, Rothacher tied research to forest productivity.[26]

At the close of the 1950s, the Pacific Northwest Research Station published a guide to the H. J. Andrews Experimental Forest describing its purpose and research programs. The research station and the Willamette National Forest, in cooperation with the School of Forestry at Oregon State College, supervised investigations on the Lookout Creek drainage. After seven years of harvests and constructing nearly sixty miles of all-weather roads, loggers had clear-cut 8.2 percent of the 15,800-acre forest. The guide noted that those activities had increased turbidity in streams during major storms, and it provided information about silt loads in the three small watersheds following several landslides. It reported that the Oregon Cooperative Wildlife Research Unit was continuing studies of logged-off areas, observing that logging operations reduced trout populations in small streams.[27]

The seminal research project on the Andrews, those small experimental watersheds, prompted scientists to extend their investigations into complex relationships among plants, water, soil, and aquatic and animal life. Given the evolution of the investigations on the small watersheds, it is important to remember that the initial purpose of the research was to seek improved strategies for managing "old-growth Douglas-fir forests on mountain streams." Those inquiries involved harvesting and reforesting methods that had the objective of establishing "vigorous young-growth stands." The investigations emphasized how "conversion will affect water quality."[28]

Jack Rothacher, who authored the first watershed study in 1958, cited the three trapezoidal flumes installed in 1952 as the beginning for gathering data. The equipment had already enabled staff to gather five years of streamflow and precipitation data. The purpose of the investigation was to ensure that "forested watersheds produce a maximum amount of timber without detriment to the supply of useable water." The ongoing inquiry would examine "soil-water-plant relationships," sediment in streams between low and peak flow, and the effects of road construction and logging, the last involving the harvest of an entire drainage with staggered clear-cuts.[29]

The north-flowing small experimental watersheds from Lookout Ridge ranged in size from 149 to 250 acres, with an elevational descent of 3,500 feet to 2,750 feet to Lookout Creek at 1,500 feet above sea level. Old-growth Douglas-fir dominated the watersheds, with surface runoff seldom rising above streambanks even during heavy rains. The data collected from the watersheds would be evaluated under various conditions of disturbance. Watershed 2 would be left "in an undisturbed state as a check." Watershed 3 was scheduled for normal harvesting, building roads and landings and, eventually, clear-cutting 25 percent of the drainage in three clear-cut patches. After burning slash and replanting, Rothacher reported, "treatment would then be essentially complete." The continuing study was expected to focus on recovery to "near normal runoff conditions," with the findings "used to insure [a] rapid regeneration of timber." Watershed 1 was proposed as a harvesting experiment, using a skyline crane to do the logging. The key to the overall investigation would be the streamflow statistics from the three watersheds under natural conditions and then the comparative significance "of the effects of treatments applied."[30]

The contractor for the Watershed 1 logging operation was Jacob Wyssen, a German who had designed the Wyssen W-200, ten-ton skyline crane, a machine being used for the first time in the United States. The Forest Service funded the pilot project in 1961 to test the viability of using such equipment to mitigate disturbances to soil in difficult terrain. The crane was expected to haul out some thirteen million board feet of timber from Watershed 1. During the course of the logging, the Pacific Northwest Research Station would look carefully at soil disturbances and evidence of increased sediments in the stream. "All of this," the journal Forest Industries observed, "is part of the long-range watershed study being carried out on the Andrews Forest." The publication reported that a telephone system linked the machine operator with choker setters and workers on the landing. It would be an odd scene, Forest Industries reported, with a logging crew, only half of them English speakers, hearing "German over the phone and up and down the hill."[31]

Two scientists critically important to future research on the Andrews, Ted Dyrness and Jerry Franklin, both spent summers on the forest when they were students. Still an undergraduate at Oregon State College at the end of the academic year in 1957, Franklin took a summer job on the experimental forest working under Jack Rothacher caring for the small watershed gauging stations. Later, he participated in running boundary surveys on the three watersheds. Dyrness, whose academic focus was soil science, first visited the Andrews on a graduate student field trip in 1955. Although his employment with the Pacific Northwest Research Station eventually took him elsewhere, Dyrness frequently returned to the Andrews to fish Lookout Creek with Rothacher. In the summer of 1961, he traveled with Rothacher through national forests in the Pacific Northwest gathering information on soil stability. During those years, he remembers becoming thoroughly versed "on what was going on in the Andrews."[32]

Despite what might seem lively activity on the Andrews during the 1960s, Al Levno characterized it as a quiet, lonely place, with little research activity other than reading the watershed gauges. Jack Rothacher, who moved to Corvallis in 1961, fought mightily to protect the integrity of the forest from Willamette National Forest officials who wanted full access to its timber. Levno, a self-described "young green kid" when hired in 1961 as a technician, joined Dick Fredriksen, who taught him how to manage the gauging stations.

When Fredriksen left to finish his degree in Corvallis, Levno was the only person with a daily presence on the Andrews. He lived in Blue River, attending to the gauging stations, checking the instruments, and encountering few people on the Lookout Creek drainage. Although he got along well with the district ranger, he thought Forest Service people, in general, were jealous of the Andrews, wanting access "to cut those decadent old-growth trees."[33]

Jack Rothacher was caught in the midst of arm-twisting between the Blue River Ranger District and Pacific Northwest Research Station. The district insisted that harvests on the Andrews should figure in its McKenzie working circle—that is, its projected cut to meet the needs of timber-processing facilities in Eugene and Springfield. Much of the discussion involving Rothacher and the district centered on a 1959 memorandum of understanding stipulating that the volume sold on the Andrews would approximate its allowable cut of eleven million board feet. Andrews research personnel were to work closely with the district ranger and the Willamette National Forest supervisor to provide the information needed to meet the national forest's harvest plan. The memorandum conceded, however, that harvests on the Andrews "*will not* be included in the allowable cut for the McKenzie Working Circle."[34]

Despite the understanding, the district ranger continued to pressure Rothacher to keep the allowable cut in proportion to the increased harvest on the working circle. Because the experimental forest lacked a separate work plan, Rothacher wrote Jerry Dunford at the Pacific Northwest Research Station urging an increase in the cut on the Andrews. Although the experimental forest was not in the district's working circle, Dunford saw no problem increasing the allowable cut. "Our problem," he wrote, "is how much do we want to cut?" Because Andrews scientists had "wanted to open the forest for more access to possible study areas," Dunford did not object "to cutting to or even above the allowable cut," even though the sale was not related to research. Dunford believed the experimental forest "should be looking forward to 2nd growth management practices. Do we have any reason for saving old growth?" If not, focusing research on second-growth trees would better serve the interests of science and management.[35]

During this quiet interlude in the mid-1960s, there were rumors that the Forest Service would disband its experimental forests as a cost-saving measure. Those stories appeared amid the Pacific Northwest Research Station's increasing emphasis on laboratory research and the completion of the Forest Science Laboratory on the Oregon State University campus. With research centered at the university, the research station's George Meagher floated a proposal to phase out experimental forests and return them to the national forests. Meagher's initiative alarmed Franklin, Dyrness, and others who feared that such a decision would free market forces and obliterate long-term studies on the Andrews. Meagher's suggestion worsened relations between Andrews scientists and staff at the Blue River Ranger Station who wanted to end scientific research on the Lookout watershed.[36]

For the Andrews Experimental Forest, the Christmas flood of 1964 was a scientifically providential happening, triggering renewed interest in large-scale precipitation events and landslides and slumps in mountainous terrain and their effects on forestry operations. Torrential rains in western Oregon and warming temperatures melted snow in the mountains, sending a huge volume of water pouring into the McKenzie and other tributaries of the Willamette River. The downpour wreaked havoc to slide-prone hillsides, scouring roadbeds and washing away bridges. The flooding McKenzie wiped out the McKenzie River (Belknap) Covered Bridge at a location where bridges had spanned the river since 1890. Small streams on the experimental forest plunged down steep mountain slopes, dumping water into Lookout Creek at a prodigious rate. Downstream in the low-lying valleys—especially the Willamette, where flooding had been a regular occurrence—floodwaters traditionally threaded their way through braided channels in broad flood plains. Beginning in 1939, storage dams built on tributary streams had gradually brought an end to the most ravenous flooding by the late 1960s.[37] The floods of December 1964, however, were a stark reminder of nature's unpredictable force.

During that Christmas week, Al Levno and Dick Fredriksen were checking the gauging stations every three hours, reading the instruments, and taking water samples. Shortly after midnight on December 21, they were driving into the Lookout watershed when their pickup's headlights revealed a huge pile of logs and debris blocking the road. Reversing direction and anxious to leave the area, they encountered another big landslide literally filling the road. Levno's memory of that night tells the story of two men fording streams in the rainy darkness amid the pounding noises of tumbling boulders in Lookout Creek, overcoming near-drownings, and eventually finding their way to a farmhouse on the McKenzie Highway several hours later.[38] For Andrews scientists, the subsiding waters revealed instructive findings about logging practices on the three small watersheds.

The 1964 Christmas floods spurred research into large-scale storms that caused landslides and disrupted tributary streams such as Lookout Creek. Ted Dyrness's study, *Mass Soil Movements in the H. J. Andrews Experimental*

Forest, indicated that road-building rather than logging practices was responsible for most of the landslides. The torrential rains were "a real eye-opener," he observed, including taking out the Watershed 3 gauging station. Dyrness's survey, carried out in April, May, and June of 1965, catalogued forty-seven mass soil movements from the winter storms, most of them associated with road-building. Such disturbances, he concluded, accelerated the incidence of landslides during major precipitation events in slopes that were notably unstable. In keeping with the applied nature of his research, Dyrness recommended reducing road mileage, improving road designs, and perhaps using skyline or balloon logging in steep terrain. Although the floods of 1964 posed significant lines for investigations, Dyrness wondered what it would take to attract scientists to the experimental forest to do research.[39] The answer rested in the International Biological Program, a policy initiative that emerged in the early 1960s.

Dick Fredriksen, who had escaped with Al Levno from the turbulent waters of Lookout Creek in December 1964, published one of the first studies on sedimentation in streams based on data from the three gauged watersheds. In the study, published in the proceedings of a sedimentation conference in 1963, Fredriksen evaluated sediment in the three experimental watersheds up to 1959 when logging roads were built into Watershed 3. His benchmark for the assessment was the condition of the watersheds—with great stands of Douglas-fir more than four hundred years old—before experimental cuttings began in the early 1960s. The undisturbed drainages ran low and clear during the summer, with sediment gradually increasing during the rainy season, reaching concentrations of 100 parts per million during heavy precipitation. After roads were extended into Watershed 3, the first rains in the fall increased the sediment load 250 times the amount in Watershed 2, the undisturbed drainage. With the passage of time, the winter sediment continued to decrease, although remaining noticeably higher than Watershed 2. Although Fredriksen tabulated the data before the Christmas storms of 1964, Watershed 3 experienced earlier landslides, indicating that road-building "produced quantities of sediment far in excess of previous years."[40]

By the 1970s, Andrews scientists had ready access to research opportunities in relatively pristine environments across the Pacific Northwest. Some of those were designated wilderness areas after Congress passed the Wilderness Act in 1964, creating the National Wilderness Preservation System. Research Natural Areas (RNAs)—smaller expanses of land, many of them in the national forest system—were even more valuable for investigations into wide-ranging and diverse ecosystems. Although other agencies and private organizations established RNAs, the Forest Service had been doing so since the late 1920s to protect areas of biological diversity for research and monitoring. One plot frequently mentioned in Andrews research publications is the Wildcat Mountain Research Natural Area, a 1,000-acre plot established in 1968 to preserve noble fir in Oregon's western Cascades. In a Forest Service publication released in 1972, Jerry Franklin is listed as the author of many of the seventy entries describing the ecology and other attributes of RNAs.[41]

Research Natural Areas in the Pacific Northwest provide landscapes to study productivity, classify plant communities, monitor air pollution, measure air and soil temperatures, and investigate woody debris in streams and issues related to succession, as well as the effects of grazing and the rate at which forest litter decomposes. In several interesting comparative investigations, scientists have matched research findings on RNAs with studies of manipulated areas on the H. J. Andrews, Cascade Head, and Wind River Experimental Forests. Although interest in research projects on RNAs has grown, a study published in 1986 indicated that many scientists and land managers were unaware of the RNAs. The lack of funding may be one of the problems, according to the study.[42]

The story of several large wilderness areas in proximity to the Andrews Experimental Forest is a complicated one for Andrews researchers. Although Congress passed the heralded Wilderness Act in 1964, few Andrews scientists were actively involved in the push for wilderness designations, despite the fact that many of them capitalized on access to such places for research. One major controversy struck close to home for Andrews personnel—on the Horse Creek drainage, a tributary to the McKenzie River. When the Forest Service created a program of L-20 Regulations in 1929 establishing "primitive areas" on the national forest system, the agency included the Three Sisters Mountains in the designation. At the urging of Robert "Bob" Marshall, head of the Division of Recreation and Lands, the Forest Service upgraded protection for those areas in 1939, creating U Regulations, an administrative decision that included the Three Sisters Primitive Area's 191,108 acres. Marshall then convinced the Forest Service to add 55,620 acres of the Horse Creek drainage, including the adjacent French Pete Creek (a tributary to the South Fork of the McKenzie River), to the Three Sisters unit.[43]

The addition of Horse Creek to protected status was unique because, unlike the Three Sisters high country, the lower valleys grew sizable stands of timber. The Willamette National Forest, in full-production mode following the Second World War, was rapidly roading its forests and readying timber for sale. To gain access to more timber, the Forest Service removed the area west of Horse Creek from protected status in 1957. In effect, Horse Creek itself would serve as a boundary for protected status, its western half now open to road-building and timber sales. It is paradoxical that the Forest Service began taking bids on timber sales on the western half of Horse Creek in 1964, just as Congress declared the Three Sisters Wilderness part of its newly created wilderness system. The Forest Service, however, had yet to target the French Pete drainage for harvesting.[44]

The French Pete Creek Valley, soon to be a national story, extended from its confluence with the South Fork of the McKenzie River at 1,900 feet elevation, southeasterly to the top of the drainage in the Three Sisters Wilderness at about 5,000 feet. Its Douglas-fir forests were about a century old in the 1960s, regrowth after forest fires in the 1800s, providing timber that would produce high-quality lumber. Although there were no roads into French Pete, Willamette National Forest announced plans in early 1968 to sell eleven units of timber in the drainage, with a volume of 18.5 million board feet. "French Pete became a battle cry," writes Kevin Marsh, a challenge to the Forest Service and the timber industry to keep commercially valuable timber from being "put off limits to logging." A Eugene organization, Friends of the Three Sisters, initiated immediate protests, and a groundswell of opposition quickly emerged across Oregon and beyond. Brock Evans, Northwest representative of the Sierra Club, assisted Eugene citizens in organizing the Save French Pete Committee, which in turn sought the support of Fourth District congressman John Dellenbach.[45]

Mike Kerrick, the district ranger at Blue River, who would eventually provide strong support for investigations on the Andrews, drew up a modified logging plan for French Pete that did little to satisfy activists who demanded a wilderness designation for the area. Because French Pete had attracted broad media attention, the Willamette National Forest delayed timber sales there until 1969. In the succeeding months, two new organizations, the Oregon Environmental Council and, later, the Oregon Wilderness Coalition nationalized the French Pete controversy, prompting Oregon's junior US senator Bob Packwood to support a wilderness designation. After a stalemate that lasted well into the 1970s, newly elected Fourth District congressman Jim Weaver joined Senator Packwood in sponsoring legislation to add French Pete to the Three Sisters Wilderness. Finally, an omnibus wilderness bill resolving French Pete and other national forest wilderness controversies passed both houses of Congress, and President Jimmy Carter signed the Endangered American Wilderness Act in 1978.[46]

The importance of the Three Sisters Wilderness and its French Pete Creek addition is evident in the variety of research ventures linking them with the Andrews Experimental Forest. The Three Sisters Wilderness was paired with the Andrews in the 1970s as part of the United Nations Educational, Scientific, and Cultural Organization's (UNESCO) Man and the Biosphere Programme, a global effort to preserve natural areas representative of major biomes around the world. This marked the beginning of coordinating investigations involving manipulated environments in the Lookout Creek drainage with pristine landscapes in the Three Sisters Wilderness. Gordon Grant, a PhD student at Johns Hopkins University, was one who used comparative studies of Lookout Creek and French Pete Creek to assess the influence of logging on the downstream contours and forms of streambeds. French Pete Creek represented the undisturbed watershed. Grant subsequently served as a Forest Service hydrologist who carried out numerous investigations on the Andrews and elsewhere. By the 1980s, in addition to wilderness areas and RNAs, Andrews researchers were making extensive use of numerous other sites—Olympic National Park, Cascade Head Experimental Forest, Wind River Experimental Forest, Mount Rainier National Park, and central Oregon's ponderosa pine country.[47]

Scientists used the Andrews Experimental Forest and nearby RNA landscapes to develop environmental gradients across the western Cascades and beyond. Combined with lands in the wilderness system, those investigations marked the onset of the forest's outreach to broader national and international research agencies. Beginning with the International Geophysical Year (IGY), scientific communities in the industrial nations initiated a Cold War enterprise linked to the Soviet Union's launch of *Sputnik* in 1957, the first satellite to orbit the Earth. The executive board of the International Council of Scientific Unions (ICSU) appointed a committee to coordinate inquiries into geophysical observations. The United States provided more than $40 million to support the geophysics-based IGY, the bulk of the expenditure devoted to technical programs related to orbiting satellites. While the geophysical year was steeped in the physical sciences, the IGY provided a cooperative scientific model for biology and the emergence of the IBP.[48]

In the fall of 1961, the Council of Scientific Unions urged the International Union of Biological Sciences to study the potential for creating an International Biological Program. The biological group made the move the following spring under the formal subtitle, "The Biological Basis of Productivity and Human Welfare." The IBP had five sections, including the terrestrial communities section, whose mission was to study the world's major ecosystems to determine whether representative samples were protected in national parks, nature reserves, or research areas. Although the terrestrial communities section was associated with land, the organizers recognized that conservation studies needed to embrace "terrestrial, fresh-water, and marine ecosystems." With committees established in several countries, the IBP held its first general meeting in July 1964. The initiative was closely linked to human influences in the biosphere and the degradation of the environment—erosion, air and water pollution, and the huge imprint of urban sprawl, a major concern in the industrial world. The fundamental unit of study would be the ecosystem, the totality of living and nonliving components that interact through space and time. Although single features of the physical world had been the subject of previous investigations, no single ecosystem had been studied from the perspective of all its organisms.[49]

David Coleman has concluded that the IBP, which lasted from 1964 to 1974, was a strikingly successful scientific enterprise for its time—although not in terms of its original objectives. The IBP began in Great Britain and Europe, the United States joining the program only belatedly. American participants, however, became major contributors to ecosystem science, especially when IBP's intellectual successor, long-term ecological research, emerged. Frank Golley contended that "the ecosystem story is largely an American tale." Although the IBP was broad in scope, involving several areas of biology, the objective of its terrestrial communities section was "to study whole systems, such as drainage basins and landscapes through team effort."[50]

The IBP marked the growing interest in systems ecology and scientific studies of ecosystems in the United States. Frederick Smith of the University of Michigan's School of Natural Resources and director of analysis of ecosystems (1967–1969) remarked that a revolution among ecologists was taking place "and the IBP is in the middle of it." American scientists quickly moved to the forefront, taking advantage of the opportunity to do ecology in a new way. W. F. Blair, president of the Ecological Society of America and chair of the US National Committee for the IBP, led the effort to gain congressional support for the new approach to ecological investigations. In a report prepared for a committee hearing, Blair proposed collaborative, interdisciplinary research in his quest for funding, telling the committee that ecologists needed "to curb some of their traditional individuality" and learn "to work in large teams harmoniously and effectively."[51] His call for collaborative, coordinated, multidisciplinary approaches would be the mantra of Andrews scientists from their involvement in the IBP and far into the future.

The Ecological Society of America's request to fund ecosystems research was strikingly successful, beginning with an appropriation of $600,000 in 1969. Although Frederick Smith dubbed that a "negligible" sum, federal funding would increase to $3 million in 1970, $6 million in 1971, and $8.5 million in 1973. Congress channeled the funds through the National Science Foundation's Bureau of Biological and Medical Sciences. With the infusion of funds, the NSF established a new budget line for the IBP. Between 1968 and 1974, the IBP's Analysis of Ecosystems program received approximately $27 million, with three-quarters of the support designated for ecosystems studies. In a summary of United States funding for the IBP, Robert McIntosh wrote, "By the standards of traditional ecology these were very large sums and were instrumental in changing the way ecology was done."[52]

Biome studies, which grew out of the IBP's terrestrial communities section, were the most ambitious of all its programs in the United States. The biome investigations focused on studying the ecosystems of five large ecological regions in the United States—tundra, grassland, desert, coniferous forests, and eastern deciduous forests. The inquiries were to develop ecosystem models for each biome to assist in managing "the ecological systems of the planet." The ecosystem components measured included production, respiration, soil and topographic information, macro- and microclimate variables, and other elements. It was clear from the outset that the complexity of the research necessitated interdisciplinary, team participation among multiple institutions and agencies.[53]

Scientists affiliated with the Andrews Experimental Forest were well-positioned for the International Biological Program. A five-page undated file, "H. J. Andrews Experimental Forest," presented the central arguments for including the Andrews in the IBP. The document offered a general strategy for designating the Lookout Creek drainage for intensive study, investigations in converting old-growth Douglas-fir to fast-growing young stands, and road-building and logging strategies and their effects on streamflow and erosion. The narrative outlined "Small Watershed Experiment," "Geology and Soil," "Hydrology of the Experimental Watersheds," and "Effects of Logging and Road Construction on Streamflow and Sediment." The document mentions the Christmas-week flood of 1964 and concludes with a section on methods for minimizing such effects, called "Erosion and Sedimentation."[54]

Jack Rothacher prefaced the application to participate in the IBP with a letter to Frederick Smith describing the ongoing research projects on the Andrews, emphasizing the importance of the three watersheds, and highlighting the forest's "long history of hydrologic and terrestrial research." Rothacher followed with the official proposal outlining the physical parameters of the forest, stressing its significance "as representative of a wide area, environmentally and vegetationally very typical of a large part of western Oregon and Washington conifer forests." His nominating statement included a section titled, "Attributes Favorable for a Study," listing previous research featuring the Franklin and Dyrness surveys of vascular plants and Jay Gashwiler's population studies of small animals. The proposal included the watershed investigations as a significant attribute.[55]

Because of their interest in the IBP, Andrews personnel compiled lists of data involving research projects on the Lookout drainage. Among the documents were biweekly reports of logging on Watersheds 1 and 3 between 1962 and 1965. The small watershed studies were critical to attracting the attention of the National Science Foundation, the agency that was, along with the Forest Service, responsible for distributing congressional funds. The different erosion and sedimentation information for the three disparate watersheds illustrated their potential for providing data on how streams and forests functioned in disturbed landscapes and in old-growth forests. Those basic ecological compilations provided powerful evidence for including the Andrews in IBP's Coniferous Forest Biome Project.[56]

Jerry Franklin prepared the "Outline of Criteria" in 1968 for the forest's "nomination as an IBP study area." From the vantage point of fifty years later, the draft offered a description of a fledgling operation, with few scientists, a paucity of resources, and library and computer facilities located at the universities in Eugene and Corvallis. Franklin's list of personnel included three Forest Service scientists present on the forest "a major share of the time": Dyrness (soil scientist), Franklin (plant ecologist), Dick Fredriksen (water chemistry), and Jack Rothacher (hydrologist). Three other Forest Service people worked there occasionally: Jay Gashwiler (wildlife biologist), James Trappe (mycologist), and Richard Williamson (silviculturist).

Most of the research on the forest at the time addressed conventional issues related to logging activity that had taken place since the early 1950s. The list included streamflow and water temperature, the effects of logging on streamflow, the effects of logging methods on stream sedimentation, Douglas-fir moisture use, the effects of logging on the chemical quality of

Jerry Franklin spent his early career as a plant ecologist with the Pacific Northwest Research Station on the OSU campus, where he became a leading figure in the Andrews Forest's participation in the International Biological Program and later its involvement in the National Science Foundation's Long-Term Ecological Research initiative. In 1986 he moved to the University of Washington as Bloedel Professor of Ecosystem Science.

water, plant succession following logging and slash burning, and logging disturbances to the soil. At the bottom of the list was a reference to the Franklin and Dyrness 1969 publication, *Vegetation of Oregon and Washington*, and Jay Gashwiler's small-mammal studies.[57]

The inclusion of the Andrews Experimental Forest in IBP's Coniferous Forest Biome Project was complicated, involving intemperate negotiations with the University of Washington, which had submitted an application to the National Science Foundation that sought to exclude the Oregon proposal. To resolve their differences, Stanley Gessel of the University of Washington's School of Forestry invited Jerry Franklin and Richard Waring, a professor at Oregon State University since 1963, to a meeting, during which it became clear that the Washington people had the upper hand. At the same time, Waring thought Washington's initiative was weak, because it was too university-centered and lacked the expansive field investigations called for in the IBP. With the NSF threatening to scuttle the Coniferous Forest Biome Project, the Andrews people, according to Franklin, were willing to kill the UW effort "unless we were going to be part of it."[58] The outcome was a two-site proposal—the City of Seattle's Cedar River drainage and the Andrews Forest in Oregon.

Having the Andrews as part of the proposal was no simple matter, because Waring, Franklin, and others in the Oregon group had been considering OSU's McDonald Forest as a study site. Waring indicated that the Andrews had certain advantages—ongoing wildlife studies to determine seed mortality and the three gauged watersheds with information dating to the early 1950s. In contrast, the university's McDonald Forest strategy was partly market-driven, subject to periodic harvests to support forest investigations and OSU's School of Forestry. Although the Andrews was distant from Corvallis, it was a well-defined site with far greater protection for research projects.[59] The future would show that the Andrews Forest was a far superior location for the Coniferous Forest Biome Project than Washington's Cedar River because of the latter's limitations for ecosystem research.

The Andrews Experimental Forest fit well with the international criteria listed in E. M. Nicholson's handbook for inclusion in the International Biological Program. The Lookout Creek watershed provided an excellent sample of a "range of major ecological formations," including representative plants and animals important to the western slope of the Cascade Range." Lookout Creek had "detailed and well-documented research" dating to the early 1950s. The site was "associated with research institutions," Oregon State University and the Forest Service's Pacific Northwest Research Station. Moreover, the Andrews Experimental Forest's 15,800 acres was "of sufficient size to support its characteristic species." Nicholson's handbook concluded with a list of long-term studies under way in Great Britain—the effects of burning, erosion studies, investigation of wetlands, and the use of pesticides and herbicides in nature preserves—inquiries that paralleled some of the studies on Lookout Creek.[60]

The research generated with IBP funds literally revolutionized scientific activity on the experimental forest. Between the initial NSF/IBP funding and its termination in 1974, a wide-ranging number of interdisciplinary scientists plied the streams and steep slopes of the Andrews Experimental Forest. Among them was OSU fisheries professor James Hall, who introduced an aquatic component to the Coniferous Forest Biome Project, leading a team of younger scientists—Stanley Gregory and James Sedell—who formed the "Stream Team," with studies focusing on Mack Creek, a major tributary of Lookout Creek. The findings of the Stream Team and Fred Swanson's parallel investigations illustrated the importance of woody debris to healthy waterways. Swanson, a geomorphology postdoc from the University of Oregon, and the Stream Team built on the earlier research of OSU professor of engineering Henry Froehlich, who created rough maps and sketches describing the influence of wood in streams—redirecting water, creating obstructions and mini dams, widening channels, and reducing the gradient on mountain streams. Trees falling into streams seldom moved very far, the team found, contributing to storing sediment and providing feeding grounds for aquatic life and helping stabilize stream banks.[61]

Those findings ran counter to prevailing management strategies on national forests, which called for removing all wood material from streams following logging operations. The Andrews' investigations under the IBP suggested that clearing wood from waterways contributed to more frequent downstream torrents of water during extreme storms. The woody debris research had management implications on the Willamette National Forest when Mike Kerrick, the former district ranger at Blue River, returned as Willamette's supervisor in 1980. If woody material in streams was important to the health of aquatic environments, timber sales contracts would no longer require loggers to remove wood from streams. Operators could save money from what they considered an otherwise onerous obligation.[62] The requirements, first to remove and then to leave woody debris in streams, invited barbed humor directed at the Forest Service, with suggestions that the agency was befuddled and confused.

The establishment of sizable reference stands on the Andrews Forest also had its beginnings under the coniferous forest biome studies. The reference stands provided clearly defined, surveyed plant communities on the western slope of the Cascade Range, reserved areas (for future research) of diverse species at different elevations, aspect, and habitats. Glenn Hawk and colleagues compiled a six-year history of reference stands established between 1971 and 1976. During the 1971–1972 season, surveyors established nineteen reference stands as a contribution to the IBP studies. Hawk paid tribute to the Jerry Franklin and Ted Dyrness reconnaissance studies for identifying plant communities at different elevations on or near the Andrews Forest.

At the invitation of the Northwest Science Association, leaders of the International Biological Program's Coniferous Forest Biome Project organized a symposium to highlight its program at the association's forty-fifth annual meeting in Bellingham, Washington, in March 1972. Jerry Franklin, Richard Waring, and several scientists affiliated with the project participated in the event. Scott Overton, of OSU's departments of Statistics and Forest Management, addressed model structures for forest ecosystems; George Brown, from OSU's School of Forestry, spoke to hydrology modeling in the coniferous forest biome; Waring and others explained the development of a grid for classifying forest ecosystems; Richard Fredriksen discussed nutrients on the Andrews; William Denison presented the complexities of measuring biomass and structure in the tops of old-growth Douglas-fir; and James Sedell presented a paper, "Studying Streams as a Biological Unit."[63]

Jerry Franklin, associate director of the Coniferous Forest Biome Project, introduced the symposium, explaining the nature of large, integrated ecological research enterprises. "Why," he asked, "has this 'big science' effort been mounted?" His response was programmatic and related to problems confronting forest managers. The critical questions in land-management decisions went beyond issues of converting old-growth stands into fast-growing new forests to maximize production. By the early 1970s, it was clear that timber harvesting was influencing water quality and the health of aquatic communities. Franklin cited a need for new information linking "parts of the ecosystems, such as land and water" and incorporating economics, sociology, and the natural sciences. Researchers were studying the time required for ecosystems to recover following disturbances (timber harvests), he contended, and there was an urgency to those questions, because "society is unwilling to wait."[64]

Franklin believed scientists had a responsibility to meet "society's need for problem-solving information," demands that ran counter to "traditional ways of doing scientific research," which had always involved research by specific disciplines. Because interdisciplinary approaches were better suited to resolving complex questions in natural resource fields, conventional approaches were outmoded. The Coniferous Forest Biome Project, Franklin argued, was "one of these integrated, interdisciplinary efforts." The biome investigations blended nicely with ecologically oriented research programs involving scientists from many disciplines focusing on all parts of an ecosystem, compiling data, framing conceptual models, and engaging in communication with their scientific peers.[65]

The University of Washington's Stanley Gessel, director of the Coniferous Forest Biome Project, followed Franklin, emphasizing the unique character of biome research, with "groups of scientists working toward solutions, rather than individuals working alone." The present effort, he noted, originated in a proposal submitted to NSF in December 1969, identifying two principal research sites, the H. J. Andrews Experimental Forest in Oregon and the City of Seattle's Cedar River watershed. Both locations had concentrations of scientists, facilities, and rich histories of research. The biome was developing an understanding of both terrestrial and aquatic components of coniferous forests. Although the program was early in its gestation, Gessel praised the good working relationships among Oregon State University, University of Washington, and the Forest Service.[66]

One of the earliest publications under the Coniferous Forest Biome Project was a significantly revised publication of Jerry Franklin's and Ted Dyrness's 1969 book, *Vegetation of Oregon and Washington*. The new title, *Natural Vegetation of Oregon and Washington*, was published in 1973 under the auspices of the IBP. The expanded volume included fresh and updated information. Franklin and Dyrness credited the Coniferous Forest Biome Project and the Northwest Research Natural Area Committee for "filling in the gaps and providing new insights" in their recent work. The book included sections on different environmental and vegetation areas, the forest zones in the two states, and a segment on vegetation in unique habitats. Appendixes described soil types and plant species and communities.[67]

Franklin, Dyrness, and William H. Moir published an important paper in 1974 involving plot studies of characteristic forests in Oregon's western Cascades, most of them located in the H. J. Andrews Experimental Forest. Additional forested landscapes extending north to the Santiam drainage and to the South Fork of the McKenzie River were included to make the plot selections more widely representative. Issued as *Bulletin Number 4 of the Coniferous Forest Biome*, the investigation involved establishing three hundred circular plots, approximately fifty to sixty-five feet in diameter, of similar species to establish "a workable classification" of forest vegetation. The plots were located over a wide range of environmental conditions—elevation, landform, slope, and aspect. The researchers studiously avoided areas of recent natural and human-caused disturbances.[68] The plot studies documented a baseline for vegetation, classifications that future scientists would revisit in the coming decades.

During the IBP, scientists began gathering climate data on reference stands in 1972, a project that concluded in 1984. A digital data machine with sensors logged information on air and dew-point temperatures, wind speed, precipitation, and solar radiation. The daytime, nighttime, and seasonal maximum and minimum averages for this period created data for seasonal highs and lows—most precipitation fell between November and March, December was the wettest month, and data was collected on the average number of days between the last frost day in the spring and the first frost in the fall. The study produced graphs providing striking statistics on long, wet winters (December, the cloudiest month) and dry summers (July, the sunniest).[69] The averages during the twelve years provided an important starting point for measuring future years of escalating climate change.

In addition to personnel from the Pacific Northwest Research Station and Oregon State University, many other scientists visited the Andrews during the summer months through the support of the IBP. Once spare of humans, the experimental forest's limited campground spaces were crowded, as were the few motels in nearby Blue River and McKenzie Bridge. There were field tours involving people from beyond Oregon. Jerry Franklin conducted two workshops involving five-day campouts with thirty or more visitors from other institutions. Collaborators from Utah State University spent time on the Andrews, as did aquatic scientists from the University of Washington. Although in later years the names of people spending time on the Andrews can be identified from grant proposals and reports, little documented evidence survives for the years during the IBP.[70]

By the time funding for the Coniferous Forest Biome Project was winding down, the 15,800-acre forest was gaining recognition for its important ecosystem investigations. Along with two other outstanding Forest Service sites—Coweeta Hydrologic Laboratory in North Carolina, home to the deciduous forest biome, and Hubbard Brook Experimental Forest in New Hampshire's White Mountains—the Andrews was gaining credibility for research programs that would extend far into the future. Jerry Franklin, on the ground floor in gaining IBP funding, spent a year in Japan in 1971 and then joined the National Science Foundation in Washington, DC, for two years as the director of ecosystem science, a position that further advanced the prospects of the Andrews Experimental Forest. OSU scientists benefited from expanded research opportunities on the Lookout Creek watershed and related areas in the Cascades—Research Natural Areas and the Three Sisters Wilderness. Robert "Bob" Tarrant, director of the Pacific Northwest Research Station in the mid-1970s, characterized the IBP years as "the glory days of the Andrews."[71]

Interviews with Andrews scientists reveal unstinting praise for the influence of the IBP. Fisheries biologist Jim Hall emphasized the importance of research focused on land and water, the terrestrial and the aquatic. Hall's insights prompted Fred Swanson and Jim Sedell to underscore one of the major collaborative discoveries during the IBP, the importance of leaving woody debris in streams, whereas conventional forest policy was to remove debris from waterways. Hall agreed, "The whole concept of the role of woody debris was quite revolutionary."[72]

Dick Waring, a principal investigator for the Andrews during the IBP, cherished the practice of working across scientific disciplines. The decision to bring in postdoctoral scientists enabled leaders to avoid the disciplinary syndrome of graduate students who were constrained to stay within their chosen fields. Swanson was an exemplar of the postdocs who arrived under the IBP to do geologic mapping on the Andrews. With a PhD from the University of Oregon, Swanson worked his way through "geomorphic processes to woody debris," eventually participating in a variety of ecological investigations. The hallmark of the IBP was to give the Andrews unity, he contended, the opportunity for self-expression across disciplines. Art McKee, hired in 1970 to oversee IBP research projects, argues that the investigations "helped to crystallize . . . the value in group research."[73]

The findings of those scientific investigations were not immediately accepted in the offices of the Willamette National Forest. For the duration of the IBP, according to Jerry Franklin, "there was a lot of friction between the [Willamette] forest and the Andrews." To Willamette officials, science was "something dangerous," even though Andrews research was pointing out "real problems" with management practices. Similar clashes, Franklin added, were taking place on every experimental forest. Despite the foot-dragging of forest managers, Swanson believes the IBP set the stage for a future in which science would have a much larger role in management decisions. Among other experiences, the IBP interlude provided an opportunity to learn the fine art of grant writing, especially with NSF.[74]

Collectively, the biome programs made significant contributions— establishing new academic centers for ecosystem studies at the University of Georgia, Colorado State University, Utah State University, San Diego State University, and Oregon State University, where IBP programs became institutionalized and continue to the present day. Biome investigations dominated the IBP effort in the United States, and the most accomplished programs, such as the Andrews Coniferous Forest Biome, the Coweeta Deciduous Forest Biome, and Hubbard Brook carried the banner of progressive studies into the long-term ecological research inquiries of the 1980s and beyond. Although IBP programs ended in 1974, funding ecosystem investigations would continue.[75]

With the end of the IBP, the National Academy of Sciences (NAS) published a lengthy report of the first large-scale effort to analyze how ecosystems functioned. An NAS committee providing oversight of the American component of the IBP declared that its investigations contributed to advances in managing forest and water environments. Scientists participating in the IBP, according to the NAS report, applauded the interdisciplinary team strategy for studying ecosystems, an approach that "turned ecology from a descriptive science into one with predictive capabilities that will aid policy-makers in resource management." The NAS assessment also cited criticisms of the IBP—that it funded second-rate scientists and that its achievements could have been accomplished with less money. Defenders of the team approach countered that block-grant funding forced researchers to integrate their findings across scientific disciplines, procedures praised by Andrews participants.[76]

In a retrospective on the threat to close the Andrews Experimental Forest and the coming of the IBP, Jerry Franklin offered this observation: "If you're not going to lose it, you must use it." With only a handful of people frequenting the place in the mid-1960s, participating in the IBP "was absolutely critical" to the survival and prestige of the Andrews. "It's been the sky's the limit ever since," Franklin told his colleagues at a Forest Science Laboratory gathering in 1991. After the IBP, the Andrews "became one of three crown jewels for experimentation in North America and perhaps the world."[77] The Andrews was at one with Coweeta and Hubbard Brook, pioneers in ecosystem studies where investigations would continue into the next century.

By the 1970s, activities on the Andrews Forest had traveled a long way from the days when Roy Silen was packing in to a largely untrammeled world to lay out road systems in preparation for the first timber harvests. Most of the roads in today's Lookout Creek drainage were already in existence when scientists flocked to the forest during the International Biological Program. Research agendas had become more complex and sophisticated, and there was the prospect that relations between the research forest and its cooperating partners, the Blue River Ranger Station and the Willamette National Forest, would improve.

Chapter 2
The Flowering of Integrated Research

Along the Pacific Coast of northwestern America, the dominant vegetation
consists of dense forests of evergreen conifers, which clothe the landscapes
from northern California to the panhandle of Alaska.
—Richard Waring and Jerry Franklin[1]

The Andrews Forest's participation in the International Biological Program
provided the catalyst for other far-reaching research opportunities. Through
the National Science Foundation (NSF), the Andrews was chosen as one
of the sites to participate in the United Nations Educational, Scientific, and
Cultural Organization's (UNESCO) Man and the Biosphere Programme.
Of much greater significance was the NSF designating the Andrews as one
of its six original Long-Term Ecological Research (LTER) sites in 1980.
That prestigious selection placed the Andrews on equal footing with the
Coweeta Hydrologic Laboratory in North Carolina and the Hubbard Brook
Experimental Forest in New Hampshire, all three located on national forests.
In addition to those investigations, several Andrews scientists journeyed north
to Mount St. Helens shortly after the great volcanic blast on May 18, 1980. The
inquiries following that event would involve the attention of a cohort of lead-
ing scientists well into the future and the publication of numerous article and
several books, including *In the Blast Zone: Catastrophe and Renewal on Mount
St. Helens*, a contemplative series of observations, essays, and poetry.[2]

The ending of the International Biological Program also involved the
publication of important articles and edited books providing a foundation for
future ecosystem research. By any measure, an important contribution under
IBP was seasonal work in 1971 and 1972 establishing nineteen forest reference
stands, most of them on the Andrews, that were subject to natural disturbances
only. The baseline for establishing the stands was the Franklin and Dyrness
classic, *Vegetation of Oregon and Washington*, first published in 1969. Most of
the reference stands (twelve) were within the Andrews Experimental Forest;

in addition, four were set up in Watersheds 9 and 10, adjacent to the southwest boundary of the Andrews; two were in the Wildcat Mountain Natural Area; and one was located a mile west of the town of Blue River. The reference stands featured a variety of vegetative communities, different environmental settings, and distinctive gradients on the western slope of the Cascades.[3]

The purpose of reference stands was to preserve representative tree communities for measuring and monitoring tree species over long periods of time. The stands would be reference points for studying the diversity of species, density, biomass, and leaf areas. The data collected from each stand was impressive—location coordinates and tags for each tree, their vigor and growth condition, numbers of saplings, tallies of shrubs, the location of decaying trees and stumps (cut by humans), and the succession of species over many years. Reference stands served as a standard for comparisons with manipulated/managed stands, with the former having desired characteristics that would serve as scientific models that might be replicated in managed tree stands. Reference stands also preserved examples of forested landscapes under natural conditions. It is more than coincidental that the long-term monitoring of reference stands and watersheds—a staple of research on the Andrews, adjacent Research Natural Areas, Mount St. Helens, and elsewhere—reflected the suggestions of the great German geographer/naturalist Alexander von Humboldt (1769–1859), who urged the systematic monitoring of weather and climate, the circulation of ocean currents, and volcanic activity.[4]

Among other Andrews scientists, Richard Waring authored several articles addressing forest ecosystems and the characteristics of coniferous forests, and one article (with Jerry Franklin) describing the evolution of evergreen coniferous forests along the Pacific Northwest Coast. Waring's publications during the 1970s focused on the significance of understanding forest ecosystems to better sustain their productive capacity. In an internal Coniferous Forest Biome Project report in 1973, he pointed out that forest ecosystems existed in relation to other ecosystems, each influencing the other, and that terrestrial and aquatic worlds were united "in a crucial and dominating way." Because roadbuilding introduced disturbances with long-term effects, scientists needed to share their knowledge with decision-makers to advance "ecologically sound legislation for the maintenance and enhancement of forest ecosystems."[5]

In successive publications Waring emphasized the importance of large, integrated investigations in pursuing ecosystem inquiries, collaborative enterprises requiring "a special structure and a special kind of people." Waring believed the most important lessons of the Coniferous Forest Biome Project were to train scientists "to bridge the communications gap between disciplines and institutions." Rigorous integration of research was necessary, because increasing harvests on the region's national forests required "management practices that assure the maintenance of the land and adjacent water resources." For Waring, scientists had a responsibility to provide precise knowledge about the difficulties of managing the nation's forests.[6]

Richard Waring and Jerry Franklin authored an informative assessment of Pacific Northwest coniferous forests in *Science* (1979). They began with descriptions of the size, longevity, and biomass of the great conifer forests from the crest of the Cascade Range to the Pacific. Among the twenty-five conifer species west of the Cascades, many individual trees were the largest and longest-lived of their species. The authors explained the evolutionary circumstances contributing to the dominance of these massive conifers, a time in the distant paleo-botanical past when conifers existed only at higher elevations. In the late Miocene (12 to 18 million years ago), conifers populated large sections of those areas, but by the early Pleistocene (1.5 million years ago), Pacific Northwest forests were essentially those we see today. Since the Pleistocene, the region's climate of mild winters with heavy precipitation and warm and dry summers had been controlling factors. In contrast, in other temperate forest regions, precipitation was distributed year-round. Waring and Franklin added that fierce storms and typhoons were largely absent in the Northwest.[7]

The most significant physical feature of the coastal coniferous forests was their enormous accretion of biomass. The trees accumulated vast amounts of biomass because of their dominance and density in forest stands, rather than existing as separate, widely spaced individuals. Although northern California's redwoods were the outstanding species for biomass accumulation in the region, Douglas-fir forests also surpassed other temperate forests in biomass. The volume of biomass in Douglas-fir, noble fir, and western hemlock forests was significant even in aging stands with dead standing trees and rotting wood on the ground. Accumulations of biomass in Northwest coniferous forests, Waring and Franklin concluded, "are the rule rather than the exception."[8]

Participating in Oregon State University's Fortieth Annual Biology Colloquium in 1980, Franklin and Waring (their authorship reversed) emphasized again the research findings scientists had made in the previous decade about biomass accumulation and the productivity of Northwest forests. Investigations under the Coniferous Forest Biome Project revealed the high rate of adaptability of evergreen forests to the marine climate—moisture, temperature, and the availability of nutrients. The dominating conifers, they believed, "appear to be an evolutionary response to climate, with cool, wet winters and warm, dry summers." Recent studies of forest ecosystems had produced significant new understandings of Northwest forests—how they had adapted to environmental conditions and natural and human disturbances. Those findings, the authors contended, should contribute to improved management strategies for the region's forests.[9]

As the H. J. Andrews Experimental Forest's participation in the IBP Coniferous Forest Biome Project was winding down, Forest Service officials in Washington, DC, were involved with the UNESCO initiative to establish a network of international biosphere reserves around the globe. The biosphere reserve idea originated in a UNESCO gathering in Paris, dubbed the "Biosphere Conference," wherein nation-states would act to protect examples of major biotic systems. Scientists from more than thirty countries drew up a plan in 1969 to establish a network of worldwide protected areas, and in 1971 UNESCO established the Man and the Biosphere Programme (MAB), "an interdisciplinary, . . . intergovernmental program of problem-based research and action focused on human-environment interactions." The concept embraced conserving the diversity and viability of biotic communities, providing sites for environmental and ecological research, creating baseline studies, and training staff and educating the public. Writing for *Environmental Conservation* in 1982, Michel Batisse noted that "long-term conservation is—and must remain—the primary function of biosphere reserves."[10]

The character of biosphere reserves differed depending on natural conditions, with areas representing biomes, unique places with exceptional natural features, and human-modified landscapes. Jerry Franklin, serving a two-year stint as director of ecosystems studies with the National Science Foundation in Washington, DC, headed the US Man and the Biosphere Programme. Franklin rephrased the program's guidelines for distribution to the US Forest Service: (1) to "conserve for present and future use the diversity and integrity of biotic communities"; (2) "to provide areas for ecological and environmental research"; and (3) "to provide facilities for education and training." The core of the program, according to Franklin, would preserve "natural areas representative of the major biomes or biotic divisions of the world." The evaluations of the United States committee would give equal consideration to conservation and research. The sites selected, however, should serve as superlative representations of a biome.[11]

The common thread linking the sites would be reserves and managed areas with shared biotic features. The biotic areas chosen should be "superlative examples of ecosystems found in a province." The United States committee cited two criteria in selecting biosphere reserves: experimental sites with "long histories of ecological research and monitoring," and large natural areas with a "limited history of research and monitoring." By October 1974, Franklin's committee had designated twenty biosphere reserves in the United States, six of them on national forests and the remainder on lands administered by the Bureau of Land Management. Two of the biosphere reserves were in Oregon, the 15,800-acre H. J. Andrews Experimental Forest and the nearby 200,000-acre Three Sisters Wilderness.[12]

UNESCO's ambition for the Man and the Biosphere Programme reflected scientific strategies carried over from the International Biological Program to protect major global biomes for study and to monitor environmental change. The UNESCO project, in which ecosystems served as research units, would provide training to improve the management of natural resources and land-use practices. It soon became apparent to many participants, especially in the developing world, that the IBP model was restricted to scientific issues and fell short of addressing immediate problems. Critics argued that the conventional IBP route would impede research that would contribute to bettering the human condition. An MAB task force in 1974 recommended the adoption of a "human use system" as a more appropriate spatial unit that would embrace a variety of economic systems. Writing in *Human Ecology* in 1974, Andrew Vayda argued that in an age of limited funding, "the comprehensive biome studies of the International Biological Program have only limited utility for dealing with urgent local and regional problems related to land use and the environment."[13]

After leadership changes in the program, UNESCO shifted the focus of MAB to environmental issues associated with desertification, the loss of tropical rainforests, and related crises in developing countries. Consequently, UNESCO reduced funding to high-latitude temperate regions and increased support for applied scientific programs in the developing world. Those opposed to the human-oriented shift in the MAB program argued that the change lacked scientific grounding. Although biosphere reserves remained important as conservation resources in many parts of the world, the original intent of the program had shifted to support the needs of local populations in developing nations.[14]

United States participation in the biosphere venture moved quickly ahead, with forty-seven reserved units established by 1980. Some sites, such as the

Coweeta Hydrologic Laboratory in North Carolina and the H. J. Andrews Experimental Forest, both located on national forests, established research and demonstration projects for their part in the program. Political problems plagued American participants beginning in the 1980s when President Ronald Reagan's administration withdrew from UNESCO. By the mid-1990s even greater difficulties emerged when right-wing conservatives began attacking American participation in United Nations programs, including characterizing the MAB as a UN threat to control public land in the United States. Because of the association with the UN, the US House of Representatives launched an investigation to inquire whether the program was in the national interest. Charges that the MAB threatened private ownership created fears that UNESCO's agenda was out to seize people's property.[15]

That unique political environment took its toll on participating federal agencies, according to Jennifer Thomsen, who interviewed public employees. Agency personnel, especially those with the National Park Service, felt they lacked the support of leadership and therefore distanced themselves from programs such as the MAB. One respondent told Thomsen that the MAB program "has been very much influenced by how responsive the federal government is to the international aspect." Another interviewee remarked about the "baggage" associated with the MAB program, "because of the political sensitivity about the United Nations." Such reactions led some national parks and biosphere sites to remove signs and plaques indicating their association with the UNESCO program. Although the George W. Bush administration renewed United States membership in UNESCO, cabinet officials treated the MAB program as a hot potato, with the US State Department passing responsibility for MAB to the Forest Service in 2000, and the service, in turn, relinquishing its obligation in 2005. Matters remained in limbo until 2014, when the US State Department attempted to resurrect the moribund biosphere program.[16]

The long-term presence of UNESCO's MAB program on the Andrews Experimental Forest was limited, ephemeral, and existed largely in name only. When UNESCO changed the parameters of the program in the 1990s, lessening the scientific component and adding a strong social element geared toward the developing world, the MAB was no longer suited to the investigations and activities carried out on the experimental forest. Mark Schulze, the Andrews Forest director, indicates that there has been no antipathy toward the UN about their association with the biosphere program. The Andrews would have remained a biosphere reserve if the program had remained broad enough to embrace their research mission. "As MAB evolved, and the criteria changed,"

Schulze indicated, the experimental forest "became a horrible fit for the program." Through the passing years, a virtually defunct national MAB committee, however, enabled the Andrews to remain in the program with virtually no oversight.[17]

American participation in the MAB had been problematic for a long time. Unfriendly presidential administrations periodically refused to cooperate with UNESCO programs, eliminated funding, and made life difficult for the US MAB committee. The US group lost interest after the MAB moved away from its explicitly scientific beginnings. Schulze, who had become familiar with the Maya Biosphere Reserve during his time in Guatemala, knew the Andrews operation was very different. He was unaware, however, that the Andrews had not been in compliance with UN MAB criteria for many years. "If US MAB had been in tune with UN MAB," Schulze is certain that Andrews personnel would have been aware before he took the forest director's position in 2008. There were discussions about creating a McKenzie Watershed Biosphere Reserve, but the organizational effort seemed insurmountable given the brief time frame to satisfy the UN MAB. "That idea," Schulze reports, "got pushed to some undefined future date." As a consequence, the Willamette National Forest withdrew the Three Sisters and the H. J. Andrews Biosphere Reserves from the International Network of Biosphere Reserves on March 14, 2017. Of the original reserve sites, seventeen have been removed for failing to meet UNESCO's biosphere criteria.[18]

In the midst of the busy investigations on the Andrews, a series of events took place distant from the Lookout Creek drainage that would redound to the benefit of broadscale, interdisciplinary, coordinated research on national forests in the Pacific Northwest. The principal episodes that triggered changes in national forest management strategies were rooted in timber harvesting on the Bitterroot National Forest in western Montana and the Monongahela National Forest in West Virginia. In Montana, Arnold Bolle, dean of the University of Montana's forestry school, chaired a committee study that published a sharp critique of forest practices on the Bitterroot—clear-cutting and reforestation activities that bore no relation to returns on investment. Timber harvesting, Bolle's report charged, was environmentally disruptive, watersheds and wildlife were ignored, and recreation users were disregarded. In brief, the Forest Service was violating its multiple-use mandate. The report, "A University View of the Forest Service," better known as the Bolle Report, attracted widespread attention in Montana and throughout the Forest Service when it was released in 1970.[19]

The 1975 Monongahela decision—*West Virginia Division of the Izaak Walton League, Inc. v. Butz*—determined that clear-cutting violated the principles of the Forest Service's Organic Administration Act (under the formal title, Sundry Civil Appropriations Act) signed into law on June 4, 1897. The Fourth Circuit Court judges traced the management of national forests from a custodial role before the Second World War to the postwar era when "the Forest Service quickly changed from custodian to a production agency." Timber industry executives viewed the legal suit against clear-cutting in the national forests with alarm and acted with a sense of urgency to rectify the ban. With the influence of powerful senators and representatives from timber-producing states, Congress passed the National Forest Management Act (NFMA) in 1976. The Bolle Report, Charles Wilkinson argues, was important to congressional debates on the bill, including criticisms of below-cost timber sales, and issues of maintaining watershed health, sustaining wildlife species, and protecting recreational use. The NFMA, in Wilkinson's view, was "a well-written statute and one that struck the best level of consensus."[20] The new law would prove critical to future research on the Andrews Experimental Forest.

The management act required every national forest to draft plans that included input from biologists, hydrologists, ecologists, archaeologists, and the general public. The measure required a "Committee of Scientists" to provide advice on regulatory requirements. Although the law did not prohibit clear-cutting, it included references to diversity in the national forest system and imposed limits on below-cost timber sales. Two earlier measures, the Endangered Species Act (1973) and the National Environmental Policy Act (1969) dovetailed with the NFMA to provide additional requirements for managing national forests. In the long run, the NFMA spurred the Forest Service to listen to biologists and ecologists and to engage the public in the planning process. Historian Kevin Marsh argues that after passage of the Wilderness Act in 1964, debates involving Forest Service policy "became far more open to broad public participation."[21] The emerging court and legislative decisions restrained conventional timber management practices and ultimately redounded to the benefit of activities on the Andrews Forest.

The National Forest Management Act and the Endangered Species Act influenced how Andrews investigations played out in society. Jerry Franklin and Art McKee, Andrews Forest director, were involved with initiatives that would shape the future direction of NSF programs and enhance the significance of the Andrews as a long-term research site. From his NSF perch in Washington,

DC, Franklin wrote McKee that he needed his assistance to structure the administration of the Andrews program so that it would be "an example for NSF." If NSF plans materialized, they would significantly benefit the experimental forest.[22]

The first indication of NSF's new direction in scientific research appeared in a circular in late 1975 inviting applications for ecological field research. Only locations demonstrating "strong ongoing research," the circular cautioned, could expect funding. Applicants should include a brief history of their studies and a description of their physical facilities, their ongoing research and publications, and their future plans; list an advisory committee of scientists; and provide a mechanism for outreach to other researchers. Proposals should itemize the financial support requested and understand that funds would not cover items associated with individual research grants—salaries, supplies, and publications.[23]

In describing the growing interest in ecosystem research, the National Science Foundation's James Callahan pointed out that national parks, wildlife refuges, and experimental forests had lengthy histories of carrying out complex environmental inquiries. Ecosystems research accelerated, according to Callahan, when Congress passed the National Environmental Policy Act in 1969, mandating the federal government to "create and maintain conditions under which man and nature can exist in productive harmony." Following the NEPA directive, the American Institute of Biological Sciences provided the National Science Foundation with a list of private and university field stations important as significant natural areas for research. Callahan cited the NSF report released in 1977, "Long-Term Ecological Measurements," as an indication of federal interest in long-term research.[24]

Under the NSF's guidance, a group of some thirty scientists drafted a list of ecological reserves with the potential for long-term research. The proposal, dubbed Experimental Ecological Reserves (EER), would operate under the auspices of the Institute of Ecology at the University of Georgia. Its purpose, according to NSF's *Mosaic Magazine*, would "provide opportunities for long-term manipulative ecological research and to establish an ecological data base that will contribute to effective management of America's land resources." *Mosaic* cited the H. J. Andrews Experimental Forest as "one highly-qualified site considered for the EER system," praising its studies of carefully monitored mountain streams, nutrient cycling in pristine and logged forests, and "the effects of those different vegetative covers on soils, hydrology, and aquatic and small animals." Through its participation in the International Biological

Program and its ongoing ecosystem investigations, the Andrews Forest had much to offer scientists.[25]

Responding to the NSF's need for field stations, the Institute for Ecology drafted a master plan for Experimental Ecological Reserves. As he did in explaining the Man and the Biosphere Programme, Jerry Franklin describes the Experimental Ecological Reserves as "representative of a major ecosystem type, natural or man modified, and will be dedicated to experimental research with experimental modification permitted as is consistent with the maintenance of the Reserve as a long-term research base that contains unmodified control areas." The quality of the sites was critically important, according to Franklin. They should be representative of a larger biological region, researchers should have access to similar areas nearby, and landscapes should have heterogeneity. He added that "many of the best candidates are existing federal experimental reserves." Candidates for the EER system would be existing federal reserves affiliated with academic institutions. There was an additional caveat, Franklin wrote: grants would be made to universities and not to federal agencies, and funds would support only gathering baseline data and logistical support for scientists.[26]

Art McKee, with Franklin's input, prepared Oregon State University's application to establish the Andrews "as a National Field Research Facility, alias, Experimental Ecological Reserve." The submission limited its request to the "development and support of facilities" for the use of scientists. Working with Dick Waring and Logan Norris at the Pacific Northwest Research Station, Franklin sought feedback from scientists on "how the scientific value" of the Andrews could be upgraded. He cautioned that the application should be modest to avoid the charge that they were beginning a major building program. The focus should be on improving the database and services for scientists, "including additional trailer working and living space." Franklin also believed the forest needed two full-time residents—a site manager and a technician—to "make the facility work."[27]

The submission to NSF, under McKee's authorship, emphasized the experimental forest's significance as a national and international study site: the Coniferous Forest Biome Project, its designation as a Biosphere Reserve, and, more recently, the Institute of Ecology identifying its potential as a National Experimental Ecological Reserve. Because of the forest's expanding research, both the Forest Service and Oregon State University were committed to managing the site "as a national facility for ecological and environmental research." In addition to the 15,800-acre experimental forest, supplemental sites included

Art McKee, hired in 1970
as the field person for the
International Biological
Program on the Andrews
Forest, stayed on after the IBP
ended and became resident
manager under the Long-Term
Ecological Research grants
beginning in 1980. McKee
remained in that position until
he retired in 2002.

several Research Natural Areas and the Three Sisters Wilderness, a large pristine area for collateral investigations.[28]

The Andrews' proposal described the history of the forest, its extensive research accomplishments and dedication to science, current investigations, and its quest for new opportunities. McKee described the existing physical facilities and cooperating administrative units—Oregon State University, the Pacific Northwest Forest and Range Experiment Station, and Willamette National Forest. The application included a site advisory committee and an external committee of scientists from diverse backgrounds to oversee planning and its research projects.[29]

In the interim, Jerry Franklin attended a National Science Foundation long-term measurements conference in Woods Hole, Massachusetts. Scientist from many subfields of ecology met to discuss the importance of long-term measurements. The conference's summary statement was blunt: "Ecology requires long-term studies." Because ecosystem processes take place over long periods of time, current research does not allow scientists to distinguish between "long-term cycles from unidirectional changes; or anthropogenically-induced changes from natural ones." Organisms with long lives, the most significant biomasses in many ecosystems, require long-term measurement. Researchers in the United States and their European counterparts were giving little attention to long-term investigations. Too much research existed in "a temporal and spatial vacuum" and lacked a contextual framework to give meaning to investigations. The scientific community, the conferees concluded, should promote long-term measurements and case histories to better understand human pressures on "natural populations, communities, ecosystems, and the earth's entire biogeochemistry."[30]

Current funding approaches, the Woods Hole conferees observed, "obstruct, rather than promote, the continued long-term studies of good scientists and good projects." Therefore, project funding should cover research spanning longer periods of time, the results and records of which would "be longer than one person's career. Sites should be selected carefully, especially in those locations where scientists have already established credible short-term investigations. Cooperative relations between institutions should be encouraged, citing relations between the Forest Service "and the academic community at Andrews Forest, Oregon, and Hubbard Brook New Hampshire, [which] should . . . serve as an example to other agencies." NSF's Experimental Ecological Reserves provided a useful model, although "the selection of long-term measurement sites need not be restricted to the EER list." Conference participants advised that long-term research projects should not be in competition with conventional short-term ventures. Finally the conference urged establishing administrative sections to distinguish among different environments: terrestrial, freshwater, and marine.[31]

The National Science Foundation published a booklet in June 1977 surveying background information about establishing Experimental Ecological Reserves (EER). A federal commission first recommended such reserves in 1974, and NSF funded a feasibility study in 1976 that proposed a new interface for scientific research: "field facilities to ensure our long-term capability for ecological research and to produce ecosystem-level data required by both scientists and policymakers." This initiative had the potential to bridge the gap between basic and applied science through "cooperative interdisciplinary studies." The emerging EER network would establish sites in twenty-eight states, 90 percent of them on federal or university properties.[32]

The Biological Research Resources Program of the National Science Foundation designated the Andrews as a national EER facility in the late spring of 1977, with funding to extend over a period of three years. An amendment to the grant required the appointment of a ten-member national advisory committee to review research proposals and management prerogatives, to advise its development as a national field research facility, and to encourage scientists outside Oregon to conduct research at the site. The committee would make recommendations involving research and facility needs, such as collecting baseline data and acquiring trailers and buildings. Appointments to the national advisory committee would be for three years, the hosts to pay travel expenses for an annual meeting.[33]

The first year under NSF's EER grant involved a shift from coniferous forest biome research involving fieldwork at Watershed 10 and other places to understanding how ecosystems respond to disturbances, especially their recovery rates. During the second year, the Andrews requested support for short- and long-term studies of disturbances at different timescales, for landscapes logged between three and thirty-five years earlier. The research centered on chronosequence studies, long-term inquiries addressing "terrestrial production and decomposition processes, erosion processes, and aquatic production and decomposition." The overarching objectives for the Andrews proposal focused on ecosystem responses to disturbances on twelve watersheds—three undisturbed and nine that had been clear-cut at various times. Scientists would be studying recovery rates and the changing physical and biological character of ecosystems.[34]

NSF support for national EER research facilities was predicated on the belief that a network of related sites should be established before they were destroyed. Ecological research had advanced to the point that sites should join with other NSF-funded locations to compare data about their ecological findings. George Lauff and David Reichle believed that Experimental Ecological Reserves would provide, through cooperating field sites, an ecological database for future experiments. To advance this strategy, the Institute of Ecology recommended coordinating research data at EER sites "to meet both scientific needs and national government goals." The institute identified a list of research locations with the potential for ecological study over decades of time.[35]

For the National Science Foundation, the Experimental Ecological Reserves project provided a bridge between the International Biological Program and the future. In an effort to advance ecological research, NSF sponsored a series of workshops between 1977 and 1979 to build interest and rapport among ecologists. A large number of scientists participated, sharing with the agency their ideas about ecological theory and how to resolve environmental problems associated with resource management. The scientists discussed long-term research across multiple sites to develop "data management systems and site-specific reference collections." Collectively, the sites would share information and publications and provide external reviews of their work. The culmination of the initiative was an NSF flyer in late 1979 inviting applications for funding long-term ecological research.[36]

In the interim, Andrews scientists responded to recommendations from their national advisory committee (under the EER) to restructure their operations. The committee praised their data sets and cooperation with the Willamette National Forest and the Pacific Northwest Research Station. Under policy advice, however, the committee urged Andrews directors to draft a clear research statement, emphasizing its long-term implications and the collaborative nature of their investigations. The committee's most significant recommendation was to include more university representatives to broaden management and to lessen "dependence on one or two pivotal individuals." The committee also urged establishing a site committee to improve management activities on Lookout Creek.[37]

The most critical item in the national committee's report was the dire need for improving facilities "essential for the support of a viable national resource such as the one envisioned for the H. J. Andrews." Under present conditions, facilities and living arrangements were scattered from the district ranger station in Blue River to the headquarters site on lower Lookout Creek. The committee recommended developing a long-range plan to establish an infrastructure for advancing "research, training, and educational activities." Andrews leaders should pursue a "staged development" of providing buildings to handle large groups for educational purposes—an auditorium, office spaces, and housing, the facilities to be located on the experimental forest.[38]

With a copy of the national advisory committee's recommendations in hand, Richard Waring submitted a supplemental grant application to the National Science Foundation requesting funds to upgrade water and sewage facilities in readiness for surplus trailers to be purchased from the Environmental Protection Agency (bringing the number of trailer modules to nine). Trailers would serve as laboratories, offices and classrooms, and bunkhouses. Such a strategy, Waring wrote, would improve facilities to meet current requirements and allow time to make provision for "replacement of trailers for more permanent structures." To achieve those objectives, Oregon State University requested $195,750 from the National Science Foundation to prepare the facilities for occupation. Waring submitted the application in August 1978, citing it as an investment that would be a "catalyst in attracting scientists to the Andrews from more than local labs and universities."[39]

Waring's request for facilities' improvements forecast the more ambitious designs of other experimental forest leaders.

Even with NSF support to advance the database and add trailer capacity, research activities had developed far beyond the forest's physical capacity. In a plan titled "Facilities Development Plan for H. J. Andrews Experimental Forest," drafted in late 1979, Jerry Franklin and Art McKee estimated that current use of the Andrews was ten times greater than in 1970. The living and working infrastructure had been patched together over several years and was "scattered and woefully inadequate." The lack of trails and temporary overnight quarters made large sections of the forest inaccessible to many researchers. Franklin and McKee proposed, therefore, short- and long-term research and management plans as an efficient way forward.[40]

They designed the proposal as a three-point research and management strategy for fiscal year 1980: "(1) Living and Working Facilities, (2) Field Facilities, and (3) Action Schedule." Their brief included Research Natural Areas beyond Lookout Creek and listed thirty-two research projects under way at the Andrews Forest. Twenty-one investigators were from Oregon State University, with NSF supporting twelve projects and the Forest Service eight. The Canadian government and the Natural Resource Council of China were providers of some funding. The plan for living and working facilities included modest accommodations on-site for researchers, including a small house, an office, and laboratory space at the Blue River Ranger Station, and a campground and small warehouse on lower Lookout Creek. Those facilities still fell short of current research needs.[41]

Franklin and McKee recommended developing a trailer park at the headquarters on lower Lookout Creek to accommodate "all weather living and working accommodations for up to 25 people on a continuing basis" and an additional fifty people for short periods at the campground. The trailers should eventually be replaced with permanent buildings. Neither the Pacific Northwest Research Station nor Oregon State University, however, was prepared to fund the facilities at the present time. If use and programs at the Andrews continued to increase, the headquarters site would need a 6,000-square-foot office/laboratory; three apartment units, with four apartments in each; a 2,400-square-foot garage/warehouse; and an instruction hall with kitchen and storage space.[42] The proposal was partially realized years later when the first permanent lodgings were completed at the headquarters site in 1987.

Having field facilities distant from headquarters was a separate issue. Cabins in the field were important to researchers, with one structure already existing at Mack Creek. The small enclosed buildings were needed to provide overnight and emergency shelters for researchers caught in severe storms of wind, rain, and snow. Cabins were imperative for research teams sampling streams during storms. Building them in remote corners of the Andrews would reduce travel time for personnel reading stream gauges and weather instruments. Because winter travel with snow machines was slow, servicing research stations often required overnights stays distant from the headquarters complex. In addition to cabins, the proposal cited a need for small three-sided trail shelters on the Lookout drainage and on three Research Natural Areas in the Cascades.[43]

With the Andrews on the verge of becoming partner to the National Science Foundation's Long-Term Ecological Research program, a cataclysmic event— the eruption of Mount St. Helens on May 18, 1980—catapulted several Andrews scientists into investigating this greatest of all regional disturbances. For three decades scientists had been studying natural and human-caused disturbances on the Lookout Creek watershed and adjacent landscapes— forest fires, landslides, slumps, and the effects of drought and severe flood-ing in forest environments. Jerry Franklin, Ted Dyrness, Fred Swanson, Dick Fredriksen, Jim Sedell, and many others had been investigating and publish-ing findings on a wide range of terrestrial and stream disturbances. The focus of that research, however, paled in comparison with the aftermath of the Mount St. Helens eruption. The blast at 8:22 in the morning spewed lava for twelve hours, transforming more than 230 square miles of coniferous forest, meadows, and crystal-clear mountain lakes and streams into "a stark gray, ash-and-pumice-covered landscape." Successive smaller eruptions continued for another six years. Fred Swanson argues that, as the largest ecosystem research community in the region, the Andrews group believed they were obligated to return the Forest Service and National Science Foundation's investment in their work—and lend their expertise to the Mount St. Helens enterprise. "The world wanted to know what happened," Swanson indicates, "and we were poised to figure some of that out."[44]

The calamitous Mount St. Helens event provided Andrews and other scientists with a "living laboratory" for studying resilience and recovery for months and years into the future. As English professor/environmental writer Scott Slovic put it, what had once been a setting for fishing, hiking, and other outdoor adventures, had "become a place for thinkers and dreamers, a source of ideas," a resource for studying and contemplating environmental change on

Mount St. Helens. After the eruption, several scientists affiliated with the Andrews Forest visited the site, some of them establishing plots to study forest and stream responses to the eruption.

a massive scale. Initially described in the popular media as a "moonscape," the blast zone proved an exciting place to study ecological recovery following a cataclysmic event. Jerry Franklin recalled stepping out of a helicopter in the "devastated zone" and seeing multitudes of plants pushing through the volcanic ash. The eruption of 1980 and the others that followed created a mosaic of terrestrial and aquatic disturbances—areas minimally affected and others thoroughly damaged. Some areas affected were far beyond the blast zone, many miles away in eastern Washington, where prevailing winds left significant deposits of ashfall. Although the recolonization of plant communities was remarkable, reestablishing the old coniferous forests, Franklin predicted, was "going to be a long time in coming."[45]

Following the blast, microbiologists were interested in obtaining samples of the murky stew that was once beautiful Spirit Lake. Just four miles northeast of the eruption, Spirit Lake was foul-smelling, filled with logs, mud, and volcanic ash. A few weeks after Mount St. Helens exploded, stream ecologist Jim Sedell proposed dropping into the lake from a helicopter to obtain water samples—"which we found later was chemically similar to a pulp-mill lagoon." Jumping into the lake from a hovering helicopter, he explained, was necessary because of the threat to the copter from the unpredictable volcano and the extremely high rental costs of the machine. Sedell told a reporter that if you were renting a helicopter for $300 per hour, "you'd better come away with some data." The *Oregonian* reported that after Sedell peered at Spirit Lake from the hovering helicopter, he took "one small step for biology" and leaped in. He paid for the leap into the toxic bacteria-laden water, reporting that he "was sick for two or three days." He admitted later that he was caught up in "the euphoria of trying to see what was going on" and, along with others, "we did a lot of foolish things."[46]

Ten years after the eruption, the *Oregonian* published a special report assessing the recovery, especially vegetative life on the mountain coming back to life. A decade earlier there was renewed plant growth of many kinds—horsetail, clumps of grass and cattail, willows and alders already providing protection and food for birds and insects, and along the North Fork of the Toutle River, elk found nourishment among the willows. Ten years on, streamside vegetation was thriving along most waterways, although plant recovery was modest at lower-elevation lakes. Renewal of plant life at Spirit Lake was slowest of all. Within the boundaries of the Mount St. Helens Monument, nature was allowed to proceed at its own pace without human intervention. Beyond the Monument, in the Gifford Pinchot National Forest and the Weyerhaeuser 68,000-acre tree farm, reforestation work was readily apparent in greening hillsides. The Weyerhaeuser Company alone planted 18.4 million seedlings on 45,000 acres of its land.[47]

Twenty-five years after the initial eruption, Virginia Dale, Frederick Swanson, and Charles Crisafulli edited a collection of articles from scientists who had observed the changes in the disturbed zone since they first visited the site. The authors, among the first scientists and journalists to visit Mount St. Helens after the eruption, returned periodically to investigate field sites and collect data on ecological recovery. Access proved problematic in the early months, with many scientists relying on helicopters for transportation. By the time pre-eruption roads provided travel to areas adjacent to the core of the disturbed zone, scientists witnessed a landscape rapidly regreening in many places. Equally valuable were a series of interdisciplinary exchanges among scientists that broadened understandings about ecosystem responses to disturbances.[48]

Changes to the landscape were unmistakable twenty-five years after the eruption. The size, diversity, and profusion of plants were remarkable, and the numerous field investigations placed Mount St. Helens as "the most thoroughly studied volcanic eruption in the world." Observers to the complex landscape of forests, meadows, mountain lakes, and streams had at hand "a rich environment and an exemplary living laboratory for study." The scientists' collective efforts revealed the story of recovering diverse flora and fauna and, among plant communities, successional changes with the passing decades. The editors of *Ecological Responses* reported that "the lessons from Mount St. Helens have been consistent with findings from other disturbance studies, while others appear to be unique to the 1980 eruption of the volcano."[49]

Although the 1980 blast devastated a large area, ecological systems showed remarkable rates of vegetative recovery. Ten years following the eruption, plants were visible across most of the disturbed landscape, either as survivors under the winter's snowpack or from seeds windblown and in bird and animal droppings, which explained their long-distance dispersal. The results impressed scientists. Although the recovery twenty-five years on was remarkable, the survival of plant and animal life in heavily disturbed places was slow, due largely to repeated eruptions that retarded biological recovery. The incidence of plant succession on the mountain exhibited behavior commonly recognized by scientists, yet the editors of *Ecological Responses* argued that Mount St. Helens provides a prototype "of how complex succession may be and demonstrates that most models are too simplistic to account for all the variation that occurs."[50]

The Mount St. Helens eruption set in motion ongoing and wide-ranging inquiries at the volcanic site. Jerry Franklin (terrestrial), Fred Swanson (geomorphology), and Jim Sedell (aquatic) were principal research coordinators on the mountain in the first few years. Franklin saw the situation as an opportunity for Andrews personnel to take their ecosystem knowledge to a new place to see what they could accomplish. In relocating their research to Mount St. Helens, Andrews leaders had access to significant Forest Service and National Science Foundation funding. Franklin delighted in the memory: "We were off the Andrews, but we were still working together." According to Swanson, "the alliances we had built up through the Andrews were really critical in going after the Mount St. Helens' opportunity."[51]

In gatherings that have become renowned among Andrews scientists, Jerry Franklin devised the idea of a *pulse*—taking a group of scientists out of their comfort zone to a distant place for a couple of weeks to live and work intensively together. When he returned from his two years with NSF, he noticed that many of the thirty to forty people working on the Andrews "didn't know each other, didn't know what each other were doing." Franklin's first pulse was a two-week outing in 1978 on the South Fork of the Hoh River in Olympic National Park, where the Andrews group "worked and got wet" and learned about each other's research interests. The events involved more than sharing information—Franklin characterized their purpose as "team-building and for science"—and their activities enabled participants to share research and check each other's hypothetical arguments.[52]

Franklin took credit for coming up with the idea, telling interviewer Max Geier that his models were earlier ecological expeditions in which large groups of scientists would go into the field for as many as six months or even years to conduct research. His adaptation of the idea included a pulse in Sequoia National Park in 1983 to stretch "our minds and our experience and our hypotheses." Although the pulses were part team-building exercises, they were also fun, productive, refreshing, and energizing. And, he says, "it's always worked." At the initial pulse on the Hoh, Franklin realized the experience would indicate that the Andrews tradition of interdisciplinary conversations could be adapted to other settings.[53]

Fred Swanson remembers that the events on Mount St. Helens in 1981 and 1982 drew more than a hundred participants: the media, members of the United States Geological Survey (USGS), Weyerhaeuser and University of Washington personnel, and Andrews scientists. Gordon Grant, finishing his doctoral dissertation at Johns Hopkins University, arrived at the end of the big gatherings on the mountain and "missed a lot of the bonding that happened around the St. Helens experience." In conversations with Franklin, Swanson, and Stan Gregory, he recalled that "a lot of the development of ideas happened in that crucible." Other pulses on Mount St. Helens took place in 2005 and another in 2010 on the thirtieth anniversary of the May 18 eruption. Simmons Buntin, founder and editor-in-chief of *Terrain.org: A Journal of the Built & Natural Environments*, traveled with Fred Swanson as their tour guide on the 2010 commemorative pulse. Buntin and his colleagues learned about the repopulation of plant and animal life on the landscape. During his weeklong residence on the mountain, he noticed that the conversations about disturbances largely ignored human influences.[54]

Swanson had a significant part in the Mount St. Helens ventures, providing an opportunity for Andrews scientists to take advantage of alliances they had built to "go after the Mount St. Helens opportunity" and test their ideas in a new environment. Their cohort of specialists in aquatics, vegetation, and thermal environments positioned them to capitalize on the myriad research venues available on the mountain. Swanson used his contacts with the USGS to gain early access to the devastated zone. He was one among many scientists from the Andrews to spend time at Mount St. Helens because it provided the opportunity to test ideas in a new setting and bring what they had learned back to the Andrews.[55]

Jim Sedell thought the Mount St. Helens eruption presented the "chance of a lifetime" to carry on stream research. "Fred Swanson, Jerry Franklin, and I got full bore into the most colossal event we've seen in our careers." Sedell is proud of the way the Andrews cohort used their "tremendous intellectual capacity when the mountain blew" and praised the work of the water community as "amazing." Norm Anderson, part of the Andrews Stream Team, was another who ventured to Mount St. Helens to "carry on the same big program back at the Andrews." Art McKee, who viewed Mount St. Helens research from his vantage point as forest manager on Lookout Creek, believed the research experience "had a tremendous effect in broadening our perspective, rubbing our nose in research priorities associated with a disturbance of immense proportions." Many Andrews investigators spent only a month or so during the calendar year—"two weeks in the field and two weeks writing the stuff up" at Mount St. Helens; nevertheless, "it was a very interesting period." That lively and intense period began to lessen in 1983, McKee recalls, when, "bingo, the dollars vanished."[56]

Beyond the research at Mount St. Helens, Andrews personnel were busy at home responding to the National Science Foundation's late-December 1979 request for proposals, titled "A New Emphasis in Long-Term Research." Recognizing that "environmental biological research" would require more time than the conventional two-to-three-year granting periods, NSF announced that it would provide "limited support for long-term research projects" beginning in 1980. It also posited a major change in the granting procedures—the awards would be made as continuing five-year grants. NSF's Division of Environmental Biology would fund pilot studies to "initiate the collection of comparative data at a network of sites representing major biotic regions of North America," with the funds not to exceed $300,000 per year. The research areas would be (1) production; (2) dynamics of populations; (3) accumulations of organic matter; (4) inorganic substances and nutrients moving through soil, groundwater, and surface water; and (5) the frequency of disturbances.[57] Experimental forests with ongoing research and new sites were invited to apply. The criteria were daunting—continuity of leadership, the expectation of institutional support, adequate physical facilities, the integrity of the landscape, an extensive bibliography and library, scientific reference collections, data storage facilities, publications, and an external review committee. Priority would be given to sites participating in the Experimental

Ecological Reserve program. Proposals were due in the NSF offices no later than February 4, 1980.[58]

Anticipating their participation in the upcoming NSF grants, Andrews personnel and cooperating partners—the "Pacific Northwest Forest and Range Experiment Station, the Forest Service, USDA, and Oregon State University, hereinafter referred to as PNW and OSU"—signed a ten-year memorandum of understanding in March 1980. The MOU firmed up years of convoluted and confusing legal matters involving a public university and three different units of the Forest Service—the PNW Station, the Willamette National Forest, and the Blue River Ranger District. Hitherto, most NSF grants, the IBP, and EER (and the soon-to-materialize LTER) had been funded through OSU. And yet, obvious to all, the university was operating on national forest property. While participants found humor in those associations in the past, it was time to give legal clarity to all parties.[59]

With the memorandum of understanding resolved, Andrews leaders proceeded with their application to establish a long-term ecological research program on the experimental forest. Their proposal was detailed, exhaustive, and convincing in presentation. Under the signature of principal investigator Richard Waring, the submission outlined five major research components: (1) succession in Douglas-fir/western hemlock forests, (2) forest/stream interactions, (3) the dynamics of young forest stands, (4) the effects of nitrogen fixers in forest soils, and (5) studies of log decomposition. Each area reflected investigations established under the forest's participation in the Experimental Ecological Reserve initiative. Components 1 and 2 entailed measuring plots "in terrestrial and riparian/aquatic environments," while components 3 through 5 involved environmental manipulations that would address important scientific questions that could be accomplished only through long-term research.[60]

The Andrews proposal exuded confidence—that it met "all criteria for an LTER site of the first rank." It praised NSF for emphasizing long-term ecological research, because many questions in biology could be resolved only "through a sustained program of measurements." The Andrews program would add to "on-going research, monitoring, and data management efforts." In line with NSF's emphasis on long-term research, the forest would coordinate its sampling and data sets with other LTER sites, especially Hubbard Brook Experimental Forest in New Hampshire and Coweeta Hydrologic Laboratory in North Carolina. Because much of the Andrews research focused on gauged watersheds, its scientists would examine terrestrial and aquatic relations, citing Watershed 10, a logged site, and Mack Creek and Watershed 2, both representative of old-growth forests. A valuable addendum to those locations would be sample plots of logged and natural forested sites elsewhere.[61]

The Andrews application cited strong institutional commitments from Oregon State University and the Forest Service's Pacific Northwest Research Station, institutions offering years of experience and assuring continuity in research and management of the Andrews Experimental Forest. Richard Waring (OSU) and Jerry Franklin (Forest Service) would be site codirectors, and Art McKee, the resident manager, would be responsible for everyday operations. McKee would advise on the acceptability of research projects, mediate conflicts, and provide on-the-ground perspective to the codirectors and the national advisory committee. The proactive Andrews submission urged the NSF to create a national administrative structure to provide "network coordination and central planning and policy."[62]

James Callahan, who worked under Jerry Franklin during the latter's two-year stint with NSF, was the person in charge of overseeing LTER applications. Working with twenty-nine applications, NSF reviewers selected six submissions for funding in their initial evaluation, including the Andrews Experimental Forest. The review committee later chose another five from the remaining twenty-four applicants. In a 1984 article in *BioScience*, Callahan remarked that "site quality and institutional commitment served as a fine filter in selecting the 11 projects," an assessment that fit the Andrews application. There was an additional noteworthy factor to its selection as one of the original LTER sites—it was the only one west of the Rocky Mountains.[63]

For the codirectors of the Andrews Experimental Forest, the selection caused little readjustment in their research agendas. The NSF/LTER grant simply meant continuing investigations already under way through the Experimental Ecological Reserve program. In an amended proposal in April 1981 requesting $233,643, Waring restated the five components of the forest's long-term research projects, providing updates on each undertaking and progress in coordinating with other LTER sites. One of their most important responsibilities, he noted, was developing and integrating a national network of LTER sites to outline common "methods of measurements and analysis." Waring believed that exchanging research teams with other LTER locations "may be a fruitful technique for developing coordinated projects."[64]

Appended to the proposal was the Andrews national advisory committee's glowing praise for its operation, applauding the directors for "sensitive and creative" ways of managing resources. The balance between scientific and educational activities on the forest seemed appropriate, although the number of people present was approaching its limits. Any expansion would require the "development of the on-site physical plant." The committee was effusive about the camaraderie it witnessed, an observation "inspirational to all who visit H. J. Andrews and a tribute to the leadership of the co-principal investigators and the commitment of individual scientists to interdisciplinary research." The site had also made dramatic progress in staffing, and improving physical facilities, and the committee praised the memorandum of understanding involving Oregon State University, the Pacific Northwest Research Station, and Willamette National Forest. Although the principal investigators seemed disappointed in the dearth of out-of-state visitors, the forest's numbers were "high relative to that experienced at comparable facilities."[65]

Exercising due diligence in its responsibility, the national advisory committee pointed to "lingering problems—the need to develop a synthesis of all ecosystem investigations, an effort that would be helpful to other LTER sites. The memorandum of understanding between the federal agencies and the university advanced the commitments of those involved, but the committee thought the agreement provided "only a loose framework in which detailed specific decisions can be made and implemented." Although there was no reason to mistrust those obligations, Andrews collaborators would be wise to gather "written case histories of how key administrative decisions are reached," any problems that arose, how they were resolved, and the commitments made, and the commitments of each signatory to the agreement.[66]

Jerry Franklin, who authored the Andrews progress report for Year 2 of the initial LTER grant, listed the installation of permanent sample plots involving stands of various ages for long-term investigations. Researchers were pursuing small-mammal studies at various elevations and studies of invertebrates in different successional stages. Under the Year 2 work plan, personnel tagged pieces of woody debris in streams to determine their fate during flooding events. Under "Population Dynamics of Young Forest Stands," Franklin reported experiments in thinning trees at different spacings. Researchers were also gathering data from reference stands established during the International Biome Program—measuring, tagging, and mapping live and dead trees, shrubs, and other living vegetation in the plots.[67]

Franklin reviewed management issues that affected "the efficiency with which the LTER program can operate." The highlight of progress was the purchase of six additional trailers that required some remodeling and weatherization for living quarters, laboratory space, and educational uses. The forest had made significant moves to cooperate with other LTER sites, participating in joint meetings and planning efforts and sharing their expertise with others. Franklin reported liaison with the Coweeta, Pawnee, Niwot Ridge, and Konza Prairie LTER sites. He personally visited University of Colorado personnel (Niwot Ridge) to offer advice on establishing forest plots; Fred Swanson had worked with the same group on analyzing ecosystem disturbance regimes.[68]

The experimental forest's subsequent reports reflected continuing advances in LTER investigations. In Year 3 of the grant (1983), the Andrews informed NSF that "progress on related long-term ecological research is excellent and has been stimulated by the LTER Program." The successes reflected work involving the five components listed in the original LTER application—succession in Douglas-fir forests, forest/stream interactions, the effects of nitrogen fixers, the dynamics of young forest stands, and the decomposition of logs. Completing the sixth year of LTER research in 1986, principal investigator Fred Swanson reported significant work on a book providing summary findings on "conifer forest and stream ecosystems and landscapes of the Pacific Northwest." Debris flows (landslides) in February 1986 had provided opportunities for studying riparian sections of affected streams. The "movement of gypsy moth up the McKenzie River toward the Andrews Forest" presented the potential for another investigation.[69]

Some faculty reported difficulty finding financial matches for experiments in which LTER funds served as "seed money." Those problems aside, it was clear the experimental forest was becoming a leading center for ecological research. Andrews personnel had developed close working relations with LTER scientists at Hubbard Brook and Coweeta. Hubbard Brook, the Andrews, the Luquillo site in Puerto Rico, and Bonanza Creek in Alaska involved Forest Service collaboration. All of the original six LTER sites were on federal land, and four of them had taken part in the International Biological Program. A cornerstone for participating in the LTER network included human influences in shaping ecological systems—logging, grazing, fluctuating fire regimes, conversions in land use, and the quality of water flowing into rivers, lakes, and estuaries. In an article published in 2003, John Hobbie and colleagues cited investigations on the Andrews Experimental Forest for its "significant impact on logging and watershed management practices in the Northwest."[70]

While LTER investigations have studied ecosystems over longer time periods, Fred Swanson indicated that they have also provided opportunities to embrace larger spatial scales—the migration of species, the effects of climate change, and comparisons of ecosystems across multiple and varying LTER sites. Disturbances to landscapes—Oregon's Columbus Day storm of 1962, changes in river channels, torrential downpours during the winter floods of 1964 and 1996, the Mount St. Helens eruption of 1980, and many other topographic events—have been the subject of Andrews investigations from the very beginning. Those different temporal and spatial factors revealed significant human and natural influences on the places we inhabit.[71]

By the mid-1980s, the Andrews Experimental Forest was attracting national and international recognition for its advances in understanding forest ecosystems. Mark Harmon, an OSU forest ecologist, led an investigation into whether current logging practices—clear-cutting, removing woody debris from sites, and slash burning—were inhibiting the potential for returning nitrogen to logged sites. His study of decomposing logs focused on the role of decaying wood in returning nitrogen and carbon to the soil. Leaving woody debris, tops, branches, and dead trees on clear-cut areas, he suggested, would mimic natural processes. Other studies involved old-growth forests, with Jerry Franklin telling the *Oregonian*'s J. L. Mastrantonio in 1987 that much of what scientists had learned about old growth "comes from research at the Andrews." Dead and decaying snags, trees toppling to the ground or in streams were valuable to healthy ecosystems. Managed forests in the United States and Europe had been treated as agricultural crops, Franklin observed, with little attention to their sustainability over long periods of time.[72]

With its reputation a fixture in the NSF/LTER firmament, the Andrews Experimental Forest had moved beyond its partnership with the University of Washington. The documented record speaks loudly about the accomplishments of the Andrews and provides convincing evidence that Lookout Creek investigations had surpassed the achievements of their Washington counterparts. As the Coniferous Forest Biome Project was nearing its end, Richard Waring suggested that scientific research in large integrated projects required "a special structure and a special kind of people to accomplish its task." During their participation in the project, Andrews officials hired postdoctoral researchers to carry out most of the on-the-ground investigations. Postdocs—in lieu of graduate students—were more flexible, free to attend scientific gatherings on short notice, able to work year-round, and had the talents to wear different hats when necessary. Graduate students, in contrast, were discipline-bound, restricted by degree requirements, and more constrained in their time commitments. "Doing research within an academic structure," according to

Waring, "did not meet the objectives of our integrated program." The Andrews also had a solid reputation for working with other research sites.[73]

Site administrator Art McKee addressed a symposium in Missoula, Montana, in 1984, where he traced the evolution of the Andrews Experimental Forest from its early reliance on the Forest Service to its association with a cadre of scientists affiliated with a broad spectrum of agencies, universities, and graduate programs. The path to that transition was the forest's involvement in the International Biological Program's Coniferous Forest Biome Project, which contributed to its participation in multidisciplinary ecosystems studies. McKee told the audience that "interdisciplinary research projects were now the most prominent investigations taking place on the Andrews." He added that reliable monitoring was important in establishing the ecosystem components to be measured. A successful monitoring program should be large, diverse, and vigorous; researchers should have common interests; relationships should be cooperative and coordinated; and it should have a quality data bank.[74] McKee articulated clearly and succinctly the successes and breadth of Andrews research.

While scientists affiliated with the experimental forest were busy with activities on Mount St. Helens and its first long-term ecological research grant with NSF, Oregon and the Pacific Northwest were in the throes of the most severe economic downturn in the lumber industry since the Great Depression. High interest rates and a nationwide slump in national home building triggered an avalanche of mill closures in 1979 and 1980 from Coos Bay and Oakridge in Oregon, northward to Forks and the Grays Harbor district in Washington. Reflecting shuttered mills in Oakridge, the North Santiam Canyon, and other operations in Oregon's western Cascades, Willamette National Forest timber sales plummeted and then began to recover in 1984–1985. Coming out of the recession, the industry consolidated into fewer, more-centralized operations, and the survivors reopened with advanced technologies that vastly reduced labor as a factor in production.[75] Although not directly related to research on the Lookout Creek drainage, those events were part of the larger environment in which the Andrews Forest functioned.

The warp and woof of the nation's political economy near and far weighed heavily on Willamette National Forest activities. Recovering markets in the mid-1980s prompted Oregon's powerful Republican senator Mark Hatfield and influential Democratic congressman Les AuCoin to amend appropriation bills, ordering the Forest Service to increase federal timber sales in the Northwest

beyond their targeted harvest rates (allowable cut). The liberal Hatfield, rank-
ing minority member of the Senate Appropriations Committee—and a supporter of most environmental issues—departed sharply with conservationists
when it came to timber issues. Writing for the *New Yorker* at the height of
the northern spotted owl controversy in 1990, Catherine Caulfield referred
to Hatfield as "probably the strongest congressional supporter of continued
high levels of cutting from public forests." Fearing lawsuits against further
timber sales, Hatfield alienated many Oregonians when he attached riders
to appropriations bills to restrict citizen court appeals. With the support of
Washington's senator, Brock Adams, Hatfield continued the practice of over-
riding environmental laws until 1990, when Congress refused to approve the
practice.[76]

Hatfield's efforts to maximize the cut on federal forests through the
1980s paralleled research on the experimental forest showing that harvesting
old-growth timber was damaging ecosystems and reducing biodiversity on
national forests. At the same time, research indicated that declining populations of the spotted owl were likely due to diminishing old-growth forests, its
preferred habitat. Environmental issues centering on the Pacific Northwest's
major industry, logging and lumbering, generated heated political and scientific debates that tore at the region's civic and cultural life. Those controversies reverberated far beyond the Northwest, challenging the viability of the
Endangered Species Act and involving Andrews scientists in policy decisions
that reached far into the future.

Chapter Three
Going National: The Spotted Owl and Other Controversies

> *The attention garnered by the spotted owl has been due not to what it does, but where it does it—old-growth forests. Throughout most of the owl's range, its habitat requirements for nesting and foraging limit it to forests with structural characteristics that take 150 years or more to develop.*
> —Jacob Bendix and Carol Liebler[1]

Andrews scientists began playing an increasingly important role in policy-making in the late 1980s and into the 1990s, reflecting their research in the International Biological Program, their growing interest in biodiversity, and their seminal investigations in the special attributes of old-growth forests. In the midst of these activities, a young scholar, Eric Forsman, began pursuing his personal interest in bird life in the early 1970s, focusing on the northern spotted owl (*Strix occidentalis caurina*). As the decade advanced, Forsman's studies eventually dovetailed with research involving the characteristics of old-growth Douglas-fir. The joining of those investigations would eventually place images of the owl on the cover of national magazines and in the headline news. Controversies surrounding the owl would subsequently serve up raw fodder for increasingly acrimonious political debates.

In the interim, Andrews scientists wrestled with events and investigations linked to the National Forest Management Act (1976), healthy ecosystems, the intrinsic values of old-growth forests, the importance of riparian zones, the river continuum concept, and how all these played out in the region's politics. The National Science Foundation, the Forest Service, and other public and private funding sources continued to support ecosystem studies, despite the anti-science bent of President Ronald Reagan's cabinet members, Interior Secretary James Watt, and Reagan's assistant secretary of agriculture, John B. Crowell, who oversaw the Forest Service.[2]

There were ironies in the environmental politics of Oregon and Washington during the 1980s. While national forests in both states became the part of Crowell's accelerated program of cutting old-growth forests, the Northwest states made impressive gains in the creation of new wilderness areas (twenty-four in Oregon and twenty-three in Washington), and Reagan signed most of the wilderness additions into law in 1984. In the midst of the hastened pace of cutting the diminishing stands of old-growth forests, Andrews scientists went about their work, attending to permanent plot systems and reference stands on the Lookout Creek drainage and nearby research natural areas, and comparable plots on the Cascade Head Experimental Forest and Olympic, Mount Rainier, and Sequoia National Parks. Through the mid-1980s, personnel continued their long-term field investigations under the National Science Foundation's LTER program—a hundred-year thinning experiment in young Douglas-fir stands, the two-hundred-year log-decomposition research in terrestrial and stream environments, and the investigation of land-management practices and long-term productivity at designated locations.[3]

Most significant among the ongoing programs were forest management research projects—commercial thinning, understanding the importance of riparian zones, regeneration issues, soil properties in Douglas-fir stands, and growth and yield studies. In a book chapter outlining work on the Andrews Forest, Jerry Franklin and Dick Waring emphasized that "almost all major basic research projects at the HJA have proved to have management applications, even though the funding source is typically the National Science Foundation." The experimental study topics included thinning, shelterwood versus clear-cutting, tree-planting on slopes at different elevations, and "studies of the ecology of the spotted owl," of which there were several breeding pair on the Andrews. "Dispersal of fledglings has been wide," the Franklin and Waring assessment reported, "but survival has been very poor."[4]

In watershed and erosion studies, scientists were giving special attention to areas where landslides occurred. The Willamette National Forest had already made use of data from the investigations in its planning. Stream and riparian researchers were employing multiple perspectives—fish population counts, the stability of streambanks, water quality, and the overall production of streams. Franklin and Waring cited the importance of woody debris, an early area of research that continued to interest scientists as an important component of waterways. Streamside vegetation was valued for the protection it provided in reducing water temperature and erosion. The authors concluded that successful ecological restoration involved interdisciplinary, collaborative inquiries.[5]

Through the early years of research, other than converting old-growth stands to fast-growing, uniform stands of trees, Andrews scientists had paid little attention to old-growth forests. Except for Jay Gashwiler's small-mammal studies that date from 1959, there was little interest in wildlife management strategies on national forests. During the International Biological Program, scientific publications, bulletins, graduate theses, and reports mostly ignored the importance of old-growth Douglas-fir to wildlife species. Mike Kerrick, district ranger at the Blue River station from 1960 to 1970 and supervisor of the Willamette National Forest from 1980 to 1988, remembers, "If there was any interest in other animals, it was more from a damage perspective, the effects of deer mice on regeneration. It was more the nuts and bolts from a science base for how are we going to manage this land." Gashwiler's comparative investigations between 1954 and 1965 of old-growth and clear-cut areas found that red-backed voles, Douglas squirrels, and flying squirrels—animals that frequented old growth—were absent in clear-cuts.[6]

Through these years, the Forest Service itself showed little interest in old-growth trees, other than its three-decades mission to convert old-growth forests to fast-growing young stands of timber. At the onset of the 1980s, the agency and the forest industry had been harvesting old-growth trees at three times their annual growth rate. Federal harvests were an important contributor to this trend, because federal harvests were filling the demand for saw logs in the face of diminishing stands of private timber. A study released in 1982 indicated a shift from harvesting low-elevation forests in the 1940s and 1950s to higher elevation timber by the 1970s. With the passing years, that pattern of clear-cutting advanced to higher-elevation national forest and BLM timber above 2,000 feet. By the 1970s, little old-growth forest remained on private land, a matter with dire consequences for amphibians, reptiles, and mammals that thrived in regenerating natural and old-growth forests.[7]

Because of the growing interest in biological diversity and the values inherent in old-growth forests, Jerry Franklin and others published a seminal Forest Service report in 1981, *Ecological Characteristics of Old-Growth Douglas-Fir Forests*. General Technical Report 118 cited the unique character of old-growth coniferous forests, underscoring features distinguishing them from "managed and unmanaged (natural) young stands. The exemplary Douglas-fir/western hemlock old-growth forest was 350 to 750 years old and possessed a litany of traits, including ideal habitats for a broad range of invertebrates and vertebrates, notable among the latter, the northern spotted owl.[8]

The Franklin document originated in a Forest Service workshop at the Wind River Experimental Forest in February 1977 to identify the ecological nature of old-growth forests and how they differed from managed forests. Although many tracts of old growth were protected in a variety of permanent reserves, they represented only 5 percent of the original forested areas in the United States. Public foresters and scientists worried that continued harvests of old growth on the national forests was endangering the unique plant, animal, and insect communities in such stands. Old-growth forests were more than "a collection of some large, old trees," differing from managed forests in major characteristics, "large live trees, large snags, large logs on land, and large logs in streams." Trees varied in species and size, their canopies producing filtered light "accentuated by shafts of sunlight on clear days." The importance of old-growth trees beyond their timber value, Franklin concluded, was their significance for communities such as mosses, lichens, and wildlife that thrive in such environments. In a later publication, he reflected on his forestry school education, which viewed old-growth forests as "biological deserts and cellulose cemeteries," trees wasting away and needing to be replaced with fast-growing young stock.[9]

Scientific interest in old-growth forests paralleled the research of Eric Forsman, a graduate student at Oregon State University who began studying the northern spotted owl (*Strix occidentalis caurina*) in 1969–1970. With financial support from the Forest Service, Forsman began full-time research on the northern spotted owl in 1972, an inquiry that would lead to his master's degree in 1976 and PhD in 1980. The northern spotted owl, with a hitherto little-known natural history, was a medium-sized owl with nocturnal habits whose primary habitat was old-growth forests where it hunted small mammals such as flying squirrels, snowshoe hares, red tree voles, and deer mice. Forsman found that the birds nested approximately sixty feet above the ground in old-growth tree cavities, deformed limbs, and in nests vacated by other species. His master's thesis research involved monitoring eighteen spotted owl nests: the number of days nesting, the length of time in which owlets fledged, and the rate of mortality between fledging and late August (35 percent). Predation and owlets falling from nests, he speculated, were the principal causes of death.[10]

Significantly, Forsman found that, of the northern spotted owl sites he found, 95 percent were in old-growth coniferous forests. Old forests, he believed, "provided owls with large trees for nests and winter roosts, small

A 1973 photo of Eric Forsman climbing a northern spotted owl nest tree to measure the cavity and collect the remains of prey in the nest. Beginning in 1969–1970, Forsman began his career-long study of the owl, establishing convincing data to show its dependence on old-growth forests.

shaded summer roost trees, and a closed canopy." Moreover, most nests were on the north or east exposures of mountain slopes, perhaps because the stands were larger and denser than on southern and western exposures. Forsman also reported that timber harvests were scheduled on half of the owl habitats surveyed for his study. He added that when small areas (smaller than 200 acres) were clear-cut, spotted owls disappeared. Forsman was the lead author in a *Wildlife Society Bulletin* of 1977 indicating that diminished stands of old-growth forests might be responsible for the decline in spotted owl populations, a topic of growing interest to scientists. In this preliminary investigation, extremely low densities of owls in young conifer stands contrasted sharply with very high numbers in old-growth forests. The *Bulletin* closed with this observation: "We doubt . . . that there will be any population increases under extensive forest management."[11]

In his doctoral dissertation, Forsman observed that his earlier findings about the northern spotted owl's overwhelming preference for old-growth forests had become "a focal point in the controversy over the harvest of old-growth forests." Preservationists cited the owl as the primary reason for the protection of old growth, while the forest industry argued that the owls could survive in second growth as well. Although those differences might never be bridged, Forsman wanted to provide more detailed surveys about the owl's habitat requirements. For a test group, he fitted radio transmitters to eight adult spotted owls and observed the birds for approximately one year.[12]

The home ranges of owls, which sometimes overlapped, varied from 2,273 to 3,400 acres, with the extent expanding during the winter months. Because their home areas were so extensive, owls were unable to protect the perimeter, but their defenses intensified closer to the nesting tree. While 92 percent of the owls in Forsman's study foraged in old-growth forests, young forest stands were seldom visited as a food source, his research emphasizing again the importance of old-growth stands. His investigation underscored the nocturnal behavior of the spotted owl—on average, foraging began "14 minutes after sunset" and ended "21 minutes before sunrise." The birds spent nights moving from one perch to the next on the lookout for prey. Forsman's research highlighted the importance of flying squirrels in the owl's diet—"42 percent of all prey captured." His conclusion: spotted owls were specially adapted for foraging and nesting in old-growth forests.[13]

Forsman followed with another publication citing research indicating that wildlife biologists and land managers were increasingly concerned about the rapidly diminishing stands of old-growth forests. To address the issue, federal, state, and private-sector biologists conducted inventories of spotted owl populations between 1972 and 1981 in Oregon, Washington, and California. The Oregon inventory, which was the most intense, embraced approximately 50 percent of spotted owl habitats and counted more than six hundred sites. Forsman observed that counting more than six hundred spotted owl locations in Oregon was remarkable, because before 1970 there were only twenty-four reported sightings of the bird. In all three states, the vast majority of owl sites were found in old forests. The large volume of data collected in the 1982 inventory, according to Forsman, conclusively documented "the strong preference of the spotted owl for old forests."[14]

In a companion investigation involving Oregon's Coast Range, the Klamath Mountains, and the east slope of the Cascade Range, Forsman found 636 spotted owl sites, more than 90 percent of them on federal land. The inquiry involved two radio telemetry studies—on the H. J. Andrews Experimental Forest and in the Coast Range on Bureau of Land Management (BLM) land west of the town of Lorane. Although both settings were mountainous with several streams, there were striking differences in the two locations. Old-growth forests still blanketed 55 percent of the Andrews and eighty- to two-hundred-year-old trees covered another 10 percent. The remaining areas were in various stages of new growth after clear-cutting in the 1950s and 1960s. In contrast, 70 percent of the Coast Range site had been harvested in the last forty years, and the remaining old-growth stands were islands in a sea of clear-cuts (about 20 percent of the study area). Biologists from federal and state agencies assisted with the owl inventory, especially in areas where private and federal lands were intermingled.[15]

Forsman wrote about the owl's behavior, its habitat preferences, and the collection of radio telemetry data. The spotted owl locations were distributed throughout the Coast and Cascade Ranges wherever old-growth forest stands existed. Investigators sighted only two owls east of the Cascades. The singular feature about the sites of virtually all the spotted owls was "the presence of an uneven-aged multilayered canopy" of old trees. Forsman characterized the habitats as having "moderate to high numbers of large trees with broken tops, deformed limbs, and heart rot. Spotted owl nests were usually located in such trees." In the Coast Range, with only patches of old-growth stands, however, spotted owls ranged over far wider terrain in their foraging activities. The capstone to Forsman's 1984 publication cited the implications for management: "The most serious threat to the spotted owl in Oregon is the gradual elimination of its preferred habitat (old-growth and mature forests)." He recommended establishing old-growth management units on federal forests across the spotted owl's range in Oregon.[16]

Forsman's research on the northern spotted owl pushed the logging of old-growth forests to the forefront of public discussions about the Endangered Species Act. Environmental organizations, including the Portland and Seattle Audubon Society chapters, began to address the need to preserve spotted owl habitats. The societies began to work with the Sierra Club Legal Defense Fund to seek ways to bring an end to old-growth harvests on federal lands.[17]

While Andrews personnel were preparing the application for a second five-year LTER grant, James Callahan wrote Jerry Franklin regarding details about the review process and items to be highlighted in their submission. "The strengths of the Andrews LTER project do not require enumeration," he observed; however, the proposal should emphasize geomorphological processes and the landscape scale of their investigations. The experimental forest's "leadership position within the LTER network is generally acknowledged," he added, "and should be carefully tended." Submitted in April 1985, the Andrews proposal

emphasized the forest's ongoing "observations of pristine ecosystems (including a 450-year-old forest)," forest/stream investigations, the importance of
riparian vegetation, and the long-term significance of nitrogen fixers. In his
final year as principal investigator, Franklin stressed the Andrews extensive
collaboration with other LTER sites.[18]

As the Andrews renewal proposal indicated, research in old-growth forests
would proceed apace in the last half of the 1980s. The investigations of diminishing old-growth forests highlighted threats to ecological diversity, complexity in stand structure, and other ecological characteristics necessary to sustain
plant and animal life. Research on the experimental forest revealed that forests
managed for fifty- to hundred-year rotations lacked diversity in stand structure,
standing snags, and woody debris on the ground, physical attributes that sustained animal and bird life, especially the spotted owl. Wildlife biologist Jack
Ward Thomas, who would later head the Forest Service, cited "a perceptual
revolution"—changes in his field during the 1970s, when wildlife research took
on a more holistic approach toward wildlife, emphasizing the need to consider
all species and their habitats. Writing in 1982, he stressed the importance of
habitat to the survival of some species. Under the National Forest Management
Act, Thomas observed, national forest managers need to monitor "indicator
species," because they reflected the health of larger groups of species.[19]

Thomas offered a prescient argument: the 1976 legislation and the
Endangered Species Act of 1973 were important to protecting wildlife on
national forests. Managers would be required to consider all species, those
taken for food, sport, or hides, and others who were vulnerable or endangered. Land-use plans needed to be comprehensive and responsive to the new
legal requirements and should meet the "bio-political test of acceptability."
Studying habitats for single species was a more rational approach for selecting indicator species, he indicated, and when they are identified, scientists
must develop a detailed plan for habitat requirements in advancing land-use
proposals. Thomas thought wildlife biologists were better equipped today to
"participate in land-use planning than they were in 1970."[20]

Through the 1980s, Jerry Franklin continued to extend his support for
the intrinsic values of old-growth forests and issues of biodiversity. Toward
the end of the decade, he drafted a management strategy to move managers away from the agricultural/plantation model of intensive forestry. In an
address to the Society of American Foresters in 1982, he touted the benefits
of the structural and compositional diversity in old-growth forests, the messy
abundance of woody debris on the forest floor, large fallen and standing snags,

the wide range in size and age of trees, the "multilayered crown canopy."
Franklin believed "large trees, snags, and logs are the key structural features of
old growth." Although he focused on Douglas-fir forests west of the Cascade
Range, he reminded the audience that old-growth ponderosa and lodgepole
pine forests east of the Cascades differed dramatically. The same was true of
old growth in the Sitka spruce/western hemlock zone along the coast where
spruce, Alaska cedar, and western redcedar were the oldest and largest trees
providing standing snags and downed dead wood.[21]

Franklin estimated that the Douglas-fir region embraced about fifteen million acres of old growth in the early nineteenth century. Of those once-vast
stands in western Washington and Oregon, approximately one million acres
of old growth remained protected in national parks, wilderness areas, and
Research Natural Areas. Many of those reserved lands were at higher elevations that supported subalpine forests, with a lower percentage of Douglas-fir.
Those statistics, he told the society, emphasized the importance of preserving
the existing old-growth Douglas-fir as conservators of water quality and producers of soil nutrients, trees that inhibited outbreaks of diseases and insects
and reduced risks from fire. Franklin closed his address with a cautionary note:
"Old-growth forests are neither the paragon of virtue and beauty imagined by
some, nor the purposeless wastelands imagined by others."[22]

At the Andrews Experimental Forest, daily life continued to center on the
numerous Long-Term Ecological Research projects associated with National
Science Foundation grants. Amid the ongoing old-growth investigations,
Andrews site director Arthur McKee published a remarkable assessment
of the changing character of investigations on the forest and the scientists
doing research. He identified the 15,800-acre Andrews and the seven nearby
Research Natural Areas as a composite research unit. Prior to the Andrews'
participation in IBP's Coniferous Forest Biome Project in 1970, Forest Service
scientists were the only ones doing research on the Andrews. The advent of the
Coniferous Forest Biome Project produced a dramatic shift involving people
from new agencies and universities working on the Lookout Creek drainage.
In the midst of the first LTER grant, McKee indicated that university researchers were now conducting about two-thirds of the investigations and bringing
with them substantial numbers of graduate students.[23]

Following this significant move, agency and university scientists were
increasingly involved in interdisciplinary ecosystem research. Those integrated
investigations eventually led to the development of coordinated monitoring

efforts and the creation of data that would be available to scientists with different research interests. Originating under the Coniferous Forest Biome Project, the monitoring program grew in scope and organization to the point that its long-term ecological collections provided useful portraits of long-term trends in anthropogenic pollution and other environmental changes. With more data, McKee believed, "environmental impact statements would have more credibility, and regional and local land-use plans could be more effectively developed by land managers." It was important for monitoring programs to have diverse and rigorous research programs, common objectives, a cooperative spirit, reliable administrative structures, responsible strategies for collecting and storing data, and firm financial support. Of those components, the quality of research was the most important.[24]

McKee explained the large and diverse investigations on the Andrews Experimental Forest. During calendar year 1983, sixty-one academic scientists, twenty-one agency scientists, and forty-eight graduate students were operating under more than fifty different funding sources. A majority of the researchers made use of data from the Andrews monitoring files. University and agency scientists, who were measuring many factors, had merged their data, an indication of the degree to which their research interests overlapped. The Andrews monitoring program, which originated several decades ago, according to McKee, began with a sampling program and evolved over the years to a more sophisticated program. "Research interests," however, determined and shaped the monitoring program that now provides long-term ecological measurements.[25]

The multitude of investigations on the experimental forest followed McKee's prescription—diverse groups of scientists investigating multiple sites, most of them west of the Cascade Range in Oregon and Washington. One study provided data on tree mortality in old-growth Douglas-fir on the Andrews and similar stands of Sitka spruce on the Cascade Head Experimental Forest. For Cascade Head, records reveal that over a fifty-year period, annual tree mortality was relatively constant. The results for the coastal old-growth setting compared well with the Andrews, where mortality, barring a catastrophic event, appeared "to be more nearly continuous than episodic." There was, however, one striking difference in the two locations—a much smaller loss to wind in the Cascades than on the coast.[26]

Old-growth investigations continued apace during the remainder of the 1980s, with studies of mortality on permanent plots and reference stands. LTER scientists studied three different age classes of trees—recent clear-cut areas, mature forests, and old-growth stands—gathering data on different age groups, their litterfall, and small-mammal populations. The last was important for developing a long-term study of small mammals, a recommendation of the national advisory committee. The special objective was to investigate small mammals in century-old and old-growth forests. The Forest Service was also sponsoring a long-term comparative study of habitats in young, mature, and old-growth forests, and the number of small mammals in each. In one report, Jerry Franklin noted that the Andrews was the ideal place for such studies and for research on the northern spotted owl.[27]

In Fred Swanson's first year as principal investigator for LTER grants (1987), he reported progress of research in the canopies of old-growth conifers and their fitness for vertebrate habitats. He described efforts "to study the feeding and nesting behavior of northern spotted owls, and the population biology of its primary prey species, the flying squirrel." Twenty-five adult owls had been radio-collared, and some were monitored around the clock. Squirrel populations were being studied in various stages of forest growth within the home range of the spotted owls. Swanson praised the cooperation between Andrews personnel and the national forest, singling out Steve Eubanks, the ranger in charge of the Blue River Ranger District, for producing a video of Andrews researchers discussing forest and stream land-management issues.[28]

During the 1980s, facilities at the headquarters site on lower Lookout Creek had undergone dramatic improvements. A National Science Foundation grant had funded trailer purchases to serve as office, laboratory, and sleeping quarters, with an overnight capacity of about fifty-five people. Sturdy roofs were constructed over four double-wide trailers to protect against heavy snows. Repairs were also being made to water-damaged ceilings, exteriors were repainted, porches were built onto trailers, and the warehouse now had a wooden floor. With another NSF grant, the Andrews prepared to build two high-elevation cabins for shelter, especially during the winter. Another grant in 1987 funded the construction of a large building to house a walk-in cooler and room for equipment, including snowcats and snowmobiles.[29]

With old-growth and spotted owl controversies heating up across the land, principal investigator Fred Swanson reported to NSF in 1989 that Andrews investigations of vegetative studies on permanent plots and reference stands were forging ahead. To sample its extensive network of plots across Oregon, Washington, California, and Colorado, the forest hired a research assistant to oversee survey crews collecting data during the summer months and then to merge the new figures with existing data sets. Thomas

Trailer under roof at Andrews Forest headquarters to protect against heavy snows. The makeshift living arrangements at the headquarters site prompted jokes about the Ghetto in the Meadow.

Spies and Jerry Franklin had prepared a publication examining forest succession on a large number of plots in a chronosequence study. Terrestrial/aquatic investigations were continuing—monitoring streams in both clearcut areas and old-growth stands. Monitoring a large debris flow originating in a severe storm in February 1986 involved gathering physical and biological information about the disturbance. Another storm in the winter of 1989 filled the same stream section with sediment, the combined disturbances revealing abundant increases in trout and fry in the debris-flow section, compared with modest numbers above the landslide.[30]

Living facilities at the headquarters site had reached a critical stage by 1989. One of the bunkhouse trailers had been condemned and towed away, while others, all manufactured in 1971, were near condemnation. Art McKee was seeking funds to construct permanent buildings to better withstand the heavy snows of winter. The big news on the facilities front, however, was the USGS decision to fund the construction of a large debris-flow flume on the northeast corner of the headquarters location. The concrete flume would be 310 feet long, 6.6 feet wide, and 4 feet deep and would be built at a 33-degree slope that would drop ninety feet from top to bottom. It was expected that scientists and engineers from around the world would use the flume "to study the physical properties and dynamics of debris flows in large-scale field experiments.[31]

"The Flume," funded and managed by the United States Geological Survey and completed in 1990, was designed to examine the behavior of large debris flows as environmental hazards.

The prospect of coordinated studies of climate change was arranged with the Corvallis office of the Environmental Protection Agency. The cooperative venture would examine the effects of climate change on forest stands. Andrews personnel were putting together another inter-site proposal to investigate the consequences of climate change on the biology of soils and how rising temperatures might degrade the sites. The 1989 report to NSF underscored a distinctive feature of the recent past—"the very high profile regionally of old-growth forests and forest management in general." Research on the Andrews, the development of new forestry practices, and research-management partnerships "have been high profile in much of the public debate concerning management of the natural resources of the region." The public's clamor for information had necessitated numerous workshops and field tours, and contributed to newspaper articles "of both editorial nature and those reporting current events."[32] The Andrews was developing a public persona!

Fred Swanson, the principal investigator for the forest's annual report in 1989, described a looming problem with a long-term study of forest production. Researchers had spent several years taking samples in an area they planned to cut, referring to it as a "natural forest to implement the study." Since they had chosen the site and completed the sampling to prepare for the cut, public opinion, politicians, and the courts had expressed great interest in old-growth forests, the northern spotted owl's preferred habitat. Swanson wrote, perhaps tongue in cheek, that "some might argue this is a product of the long history of research at the Andrews Forest on these subjects." He added, however, that a pair of spotted owls was nesting near the study site, circumstances that prevented cutting the natural forest to launch the research project.[33]

The onset of a new decade signaled the beginning of LTER3, the forest's third five-year participation in the National Science Foundation's LTER enterprise. The filing document assured NSF that the experiments and programs in the first and second LTER programs would be carried over to the 1991–1996 granting period—climate-change investigations, streamflow studies, histories of wildfire, and geomorphic inquiries. Scientists also hoped to build models of landscape responses to natural disturbances, land-use practices (converting old growth to managed forests), and the future consequences of a warming climate. Some of those projects involved inter-site cooperation. The Andrews planned to continue data management and networking activities and be attentive to "highly relevant major issues concerning natural resource management in the Pacific Northwest." Principal investigator Swanson summarized the forest's ambitions, pointing to its increasing emphasis on "natural disturbances, land use, and climate warming on forest/stream ecosystems."[34]

The National Science Foundation's approval of Andrews participation in the LTER3 continuing grant would carry the experimental forest through fiscal year 1995. James Schindler, director of NSF's Ecosystems Studies Program, congratulated the team's effort for their award and outlined the requirements, annual progress reports ("informative but brief"), and, at the end of the granting period, a final report within three months after the grant expires. The NSF review panel praised the Andrews proposal for its "outstanding research in forest ecology" and applauded collaboration between the Forest Service and Oregon State University, a "model for such arrangements." The Andrews Forest was "clearly among the LTER leaders in data and information management, probably the best." One panelist referred to the Andrews as "the archetype for LTER research. It is also a complex, very large program of which the NSF-LTER funding is a relatively minor component."[35]

As the Andrews Experimental Forest advanced its investigations in the early 1990s, external events involving old-growth forests and the northern spotted owl, much of it rooted in research inquiries on the Andrews, exploded in the national media, revealing the sharply diverging arguments between environmental organizations and timber industry supporters. Environmental protests during the 1980s, challenging the harvest of old-growth forests (to protect the spotted owl and other species), reflected a generation of citizens who valued natural systems as something more than a source of raw materials. The increased interest in outdoor recreation in the 1960s and 1970s culminated in the 1980s with backcountry travelers appreciating forests for their solitude and their importance to the looming challenges of climate change. Environmental organizations found legal support in the National Forest Management Act requiring national forest planning to include the viability of wildlife species and their habitat requirements. The environmental community found its powerful legal grounding in the northern spotted owl and its habitat—old-growth forests.[36]

In the midst of the growing media firestorm, and aware of the importance of the region's lumber industry, Jerry Franklin proposed a middle-ground management strategy between intensive forestry and outright preservation. Franklin had been developing what was eventually dubbed "new forestry," underscoring the importance of old growth to healthy forest environments. He began proposing alternatives to clear-cutting in the mid-1980s to conserve the ecological characteristics of natural forests. His proposals recommended structural diversity over the simplicity of intensive forest management practices, urging forest managers to "maintain or enhance complexity." He advised foresters to become "ecologically prudent managers," making decisions to promote the diversity implicit in the complexity of nature. There were few absolutes in ecological forestry, but a prudent operating principle should be, "simplification is rarely beneficial."[37]

Franklin believed it was unfortunate to limit debate to dividing forest landscapes into categories for either commodity production or preservation. He proposed "a kinder gentler forestry," embracing both commodity and ecological values. "Such a 'new forestry' uses ecological principles to create managed forests superior to those created under common current forestry practices." Franklin extended his new forestry argument in a presentation at the University of British Columbia in January 1989, while signaling the importance of woody debris to healthy forest environments, including snags and downed logs. Such woody detritus was equally important to healthy stream, river, and estuary life. His strategy was to create a landscape of uneven-aged trees to provide different canopy layers for ecological purposes. Franklin compared his new forestry to "ecological engineering," an approach to cope with the looming threat of climate change. At an August meeting of some 150 economists, biologists, and environmentalists in Eugene, he mentioned the difficulty of convincing the Forest Service to revise its practices in the field—"You can't just draw a line, pass an act, and get it done." Given those hurdles, "it's probably more important than anything we've ever done." [38]

Anna Maria Gillis, then with the American Institute of Biological Sciences in Washington, DC, praised new forestry's premise for proposing to maintain forests "as complex ecosystems rather than as tree factories." Gillis recounted the political and legal scuffles involving the northern spotted owl and old-growth forests, including the Fish and Wildlife Service's declaration in June 1990 listing the owl as a threatened species. She referred to the decision in the context of Jerry Franklin's testimony to a US House committee in May that "new forestry practices would create managed forests suitable as habitats for spotted owls." Peter Morrison of the Wilderness Society told Gillis that new forestry was an untested idea and provided no assurance that spotted owls could survive in younger forests. She noted, however, that the Blue River Ranger District was already writing new forestry requirements into some of its timber sales. Gillis believed there was general agreement that forest management should "be more ecologically based." [39]

The convoluted and murky world of federal legislation, environmental impact statements (EIS), and dueling lawsuits led to Congress creating the Interagency Scientific Committee in October 1989 "to develop a scientifically credible conservation strategy for the northern spotted owl." The bird would be at the center of the committee's investigation because of its preferred habitat in old-growth forests in the Pacific Northwest. The cooperating agencies included the Forest Service (FS), Bureau of Land Management (BLM), Fish and Wildlife Service (FWS), and National Park Service (NPS), all signatories to the committee's charter under Section 318 of Public Law 101-121. The committee members were Jack Ward Thomas (chair), Eric Forsman (FS), E. Charles Meslow (FWS), Barry B. Noon (FS), Jared Verner (FS), and Butch Olendorff (BLM). Forest Service members dominated, because 74 percent of the spotted owls' habitat of 7.1 million acres was on national forest lands. BLM managed 12 percent, the National Park Service managed 8 percent, and the Fish and Wildlife Service had two refuges in Oregon and two in Washington with small areas suitable for the owl. [40]

The three American states harboring spotted owl habitat had already taken action on the bird's imperiled status. Washington had listed the owl as "endangered," Oregon as "threatened," and the California Department of Fish and Game as "a species of special concern." Federal agencies had also addressed the troubled owl. The Forest Service referred to the bird as a "sensitive species" and an "indicator species," the BLM as "a special status species," and the Fish and Wildlife Service, which had dismissed previous requests to list the owl, was considering listing it as a "threatened species." The laws and court decisions regulating federal, state, and private lands in the Pacific Northwest reflected increasing social alarm for the environmental health of all plant and animal species and their habitats. "What seems to be emerging," the Interagency Scientific Committee reported, "is a growing concern with retaining and enhancing what scientists call 'biodiversity.'"[41]

The committee reviewed multiple studies about spotted owl habitat and determined that the bird preferred canopy closures, large overstory trees with significant cavities and broken tops, large snags, and lots of woody debris on the forest floor. Moreover, when spotted owls were found in younger forests, their roosts were usually in nearby older trees. In the committee's lengthy 427-page document, including twenty-two appendixes, it proposed conservation initiatives to protect the owl's habitat to ensure its survival, and follow-up research to determine whether the strategy was working. The committee proposed establishing large (50,000+ acres) habitat conservation areas (HCAs) for a minimum of twenty pairs of owls on appropriate federal lands and locating the HCAs no farther than twelve miles apart. The Coast Range presented problems, however, because of the paucity of old growth harboring twenty pairs of owls. Except for existing timber sales, logging would be prohibited in protected areas.[42]

A later review of the Interagency Scientific Committee's report praised it as "one of the most complete assessments of a species to date." In the long history of public land controversies in the United States, few resource-management issues had involved such intense debates as those surrounding old-growth forests in the Pacific Northwest. Concern for the northern spotted owl triggered a greater interest in ecosystems, and it generated opposition from politicians, the timber industry, and economists who decried curtailing timber harvests on federal forests and the adverse effects of job losses. The legal clashes over the owl began in 1987 when environmental organizations petitioned the Fish and Wildlife Service to declare the bird endangered. The agency denied the request but sought an agreement with the Forest Service and BLM to "protect spotted owls informally." When environmentalists challenged the Fish and Wildlife decision, the agency followed with its June 1990 decisions listing the owl as a "threatened species."[43]

Amid fears that legal declarations would severely restrict logging on federal forests, Oregon senator Mark Hatfield and Washington senator Brock Adams successfully attached a rider to an appropriations bill requiring the Forest Service and BLM to offer a minimum of 7.7 billion board feet of timber for sale in 1989–1990, a small reduction from previous sales. Known widely as the Northwest Timber Compromise, the legislation included Section 318—good for one year—requiring the Forest Service and BLM to minimize fragmenting ecologically important old-growth forests. The Oregon Natural Resources Council later labeled the move "the Rider from Hell." In the meantime, Circuit Court Judge William Dwyer considered the requirements in Section 318 sufficient to lift (in November 1989) an earlier injunction against federal timber harvests. Environmental organizations then appealed Dwyer's decision to the Ninth Circuit Court of Appeals, which overturned his verdict because Section 318 unconstitutionally limited review of timber sales.[44]

The listing of the northern spotted owl and the simultaneous release of the Interagency Scientific Committee's report did little to lower the heated rhetoric. In the midst of those swirling events, the prestigious National Research Council published a report, *Forestry Research: A Mandate for Change*, that struck directly at the contentious questions before the public—"how forestry research benefits society" and the "research needs of those who use forests." Authored by the Committee on Forestry Research, the fifteen-member interdisciplinary group included Jerry Franklin and James Sedell. Its report acknowledged at the outset that "the relationship between forestry and agriculture in this country (and elsewhere) needed to be enhanced and improved." The committee's "Executive Summary" parroted in large part the Andrews Experimental Forest's playbook, listing research priorities that should be strengthened: "the biology of forest organisms," "ecosystem function and management," "human-forest interactions," and "wood as raw material." The committee further underscored the purposes of forestry research: "understanding the basic biology and ecology of forests," "developing information to sustain productivity of forests as well as protect their inherent biological diversity," and "designing and implementing landscape-level and other large-scale, long-term experiments."[45]

In several pages of conclusions and recommendations, the Committee on Forest Research cited the need for more funding, strengthening graduate education in forestry, and establishing centers focusing on social benefits of forestry. Forestry research needed a new paradigm, emphasizing the environment, that focused on landscapes over long periods of time. The forest research community needed to be more inclusive, inviting scientists from other fields to investigate sustainable development, forests and global carbon balance, climate change, and promoting biological diversity. Forest scientists had an obligation to improve their communication with the public and natural resource industries. The summary of the report urged a radical change in forest research, emphasizing global warming, desertification, acid rain, and related environmental issues.[46]

By the late 1980s, the spotted owl/old-growth issue had been nationalized to the point that Forest Service rank and file were pressuring leadership to pay more attention to land and stewardship values. Writing from Seattle in March 1990, *New York Times* journalist Timothy Egan reported that the Forest Service had made record harvests from national forests in the last three years, "more than 12 billion board feet, or triple the number of a generation ago." Moreover, half of national forest logging had taken place in the Pacific Northwest, where the agency had cut "the oldest and biggest trees first." A research scientist loosely affiliated with the Andrews, Chris Maser, thought the Forest Service was treating national forests as "European-style plantations that are really nothing more than economic tree farms." Jerry Franklin told Egan that the nation's federal forests "should be more than agricultural lands with a slow-maturing crop." Forest Service chief Dale Robertson, who acknowledged the importance of Franklin's work, declared that the agency would recognize "the importance of old-growth on a national level." In a sharp critique of Forest Service policy, Richard Behan, author of *Plundered Province: Capitalism, Politics, and the Fate of the Federal Lands*, reported that several forest supervisors told Robertson in 1989 that harvests on national forests were "unrealistically high."[47]

In an October 1991 editorial, the *Times* continued its inquisition of federal forestry policy, citing Circuit Court Judge William's Dwyer's ban on logging in old-growth forests as evidence that the government had failed to effectively protect spotted owls. Despite the requirements of the National Forest Management Act, that forests should be treated as "valuable ecosystems," "the trees kept falling." The *Times* faulted President George H. W. Bush's aides for refusing to set aside sufficient acres of old-growth habitat for the owl. "That's what set Judge Dwyer off," and he found it incredible that the most powerful nation on Earth failed to protect its "irreplaceable old-growth forests and ease the pain for workers, families, and communities."[48]

Large-circulation urban newspapers in the Pacific Northwest made similar arguments, asking for political solutions to the old-growth dilemma. *Seattle*

Times writer Bill Dietrich captured the public's spirit toward the region's public forests, observing that Washington and Oregon were "wrestling with the
fate of a unique ecosystem: the ancient old-growth forests." Tree farms, he
wrote, were different "from nature's original design." The Northwest and its
timber industry had prospered from the once seemingly endless blanket of
conifers, but the replanted forest was "far different biologically than the original ecosystem." Attending a timber conference in Newport, Oregon, Dietrich
reported differences between Jerry Franklin, who thanked his stars that there
were still significant stands of old-growth forests, and industrial foresters, who
criticized new forestry. Franklin agreed that more research was needed but
insisted that logging methods could be modified to preserve characteristics of
natural forests while producing logs for consumers.[49]

The Northwest's forestry controversies spilled over into election-year politics when Oregon's senator, Mark Hatfield, the architect of maximizing federal
timber harvests during the 1980s, suddenly found himself in the midst of a difficult reelection battle in the fall of 1990. The popular Hatfield, who was running for a sixth term, had never been seriously challenged since winning his
Senate seat in 1966. Insurgent Democratic Party candidate Harry Lonsdale,
who operated a small high-tech firm in Bend, attacked the scholarly Hatfield as
a tool of the lumber industry. An outspoken, free-spirited outsider, Lonsdale
criticized the Washington establishment for "robbing us blind," and said that
Oregon's senior senator was "doing nothing to stop them." Hatfield, whose
campaigns traditionally involved quiet, small, town hall gatherings, suddenly
found himself down in the polls. After screening Lonsdale's television ads,
Hatfield told his adviser, Ron Schmidt, this was his first experience with "the
virus of negative campaigning." The *New York Times* took note: "One of the
institutions of the Senate . . . was faced with the toughest fight of his career."
Although the candidates differed on issues such as abortion, Lonsdale's
purpose for entering the race was to emphasize the need for more rigorous
restrictions on logging old-growth forests. Vastly outspending Lonsdale and
resorting to negative campaign ads himself, Hatfield won a fifth term in the
Senate, prevailing 53.7 percent to 46.2 percent.[50]

Northwest forestry issues rolled over into the presidential election year
of 1992, which found President George Bush in a tight reelection campaign
with Bill Clinton and Ross Perot. The president, who visited Washington State
in mid-September, told an audience that the Endangered Species Act should
be amended to "put people ahead of owls." In a mid-September editorial,
"Don't Blame the Owl," the *New York Times* criticized Bush for his willingness

Mark Harmon, an OSU professor of forest ecology, conducted long-term studies of log decomposition and was the lead principal investigator for Long-Term Ecological Research grants from 2002 to 2008.

to weaken the Endangered Species Act and for telling loggers that sacrificing spotted owls would save their jobs. Timber jobs were already moving to new forestry plantations in the South, the *Times* reported, and automation in the mills and woods of the Pacific Northwest was reducing labor as a factor in production. Moreover, Bush had done little to create a protection plan for the owl; neither had he addressed support for retraining out-of-work loggers and millworkers. A few days later, a letter to the *Times* disparaged clear-cuts that were destroying ecosystems. People should visit "the Pacific Northwest and see for yourself—see what we'll leave generations to come if we heed those who serve only themselves."[51]

Beyond the political jousting, the media continued to report the latest research about old-growth forests, much of it carried out on the Andrews Experimental Forest. An Associated Press release in early 1990 reported Mark Harmon's findings that cutting old-growth trees released "millions of tons of carbon dioxide into the atmosphere," thereby contributing to greenhouse gas emissions. Large amounts of carbon were stored in old-growth trees, Harmon told a reporter, far beyond the capacity of young forests. He and his colleagues were worried that decisions might be "based on inaccurate armchair ecology." In an interview with a Eugene *Register-Guard* reporter about the northern spotted owl, Harmon characterized the bird as "only a tiny piece of a rich mosaic of plants and animals that make up an old-growth forest." Protecting the owl, however, was an important move toward preserving a significant part of the region's landscape. For Harmon and other scientists, the owl was one representative in a larger cast of birds, mammals, insects, plants, and microorganisms in old-growth ecosystems.[52]

Fred Swanson thought the intense focus on the owl muddied and simplified a complex set of circumstances. The real issue was a "land allocation battle," with polarized factions wanting to either lock up old-growth forests or turn them into board feet of lumber. "We'd be best off as a society if we could look at a whole system and together sort out how to deal with it." Dave Wilcove, senior ecologist with the Wilderness Society, believed listing the owl as a threatened species elevated Pacific Northwest forests "as a national, even global treasure." The United States was dealing with remnants of the once-vast forests that covered much of the continent. He thought the nation's intellectual and financial resources should be committed to protecting the remaining old growth, otherwise "we cannot expect any nation ... to protect their natural resources."[53]

Regional and national reporters praised the H. J. Andrews Experimental Forest for providing federal officials with the research tools to revise management strategies on the nation's public forests. Salem's *Statesman Journal* commended Andrews scientists for keeping a close eye on some of the world's oldest trees, mapping, tagging, and measuring changes through the years. Those investigations, Swanson told the newspaper, were influencing Forest Service management strategies. "Understanding the complexity of old growth has played a pivotal role" in framing important issues in today's news. Moreover, he argued, old-growth studies at the Andrews had contributed to shifting public attitudes, prompting citizens to visit old-growth forests in greater numbers.[54]

The increased importance of ecosystem perspectives in forestry science prompted Forest Service leaders to announce a "new perspectives" initiative in 1990 featuring ecological strategies that highlighted stewardship efforts to sustain diverse, strong, productive ecosystems in managing federal forests. The agency's Hal Salwasser announced that the policy reflected new scientific information about the importance of "diversity and complexity in sustaining ecological systems," recognizing changing public values, and incorporating public concern about clear-cutting and the use of herbicides. Salwasser cited Jerry Franklin's new forestry as a representative example of the new-perspectives approach. Writing in Canada's major forestry publication, the *Forestry Chronicle*, J. Stan Rowe was skeptical, declaring that the industrial paradigm had guided forest practices for more than a century. "Is this the expression of a new paradigm—the 'New Forestry'—or is it an example of paradigm tinkering?"[55]

Despite the widespread acclaim for new forestry, Oregon State University engineering professor William Atkinson emerged as a sharp critic of the proposal. Addressing the annual meeting of the Oregon chapter of the Society of American Foresters in Eugene in May 1990, Atkinson excoriated new forestry as obsolete, as "hobby silviculture," and as technically "a disaster." When Jerry Franklin first proposed the new forestry approach, he regarded it as a middle ground between the lock-it-up crowd and intensive forestry practices. That approach, Atkinson argued, gave the high ground to preservationists and would lead to the loss of productive land, "the locking up of our timberlands." Tinkering with the commercial supply of logs available to mills was too risky for "hobby silviculture." Atkinson cited dire statistics—a 55 percent reduction in the availability of federal timber from the 1983–1987 average. New forestry, he charged, was actually old forestry, reflecting management policies from the late 1930s; it was too expensive and would result in decreasing production. New forestry also contributed to sloppy plantations filled with brush and weeds, killers of shade-intolerant Douglas-fir.[56]

Atkinson continued his attacks on new forestry when he joined a Franklin-organized tour of the H. J. Andrews Experimental Forest on August 12, 1990. Franklin told the assembled guests that the issue was not whether clear-cutting was right or wrong, it was simply that society preferred different forest management practices. The objective of new forestry was to manage forests in ways that benefited wildlife and enhanced biodiversity. To achieve those ends, it was necessary to mimic natural processes to promote a healthy new forest. Atkinson interrupted, telling the group that new forestry was being marketed "like a bar of soap," and to attack clear-cutting: "We're getting New Forestry by decree, by dogma. We don't have Chairman Mao; we've got Jerry." An angry Franklin countered: "If you want to, you can argue ecological values aren't important enough to interfere with economic values." He pointed out that Atkinson had missed the point: "Let's not argue about the train that's coming down the track."[57]

Charles Philpot, director of the Pacific Northwest Research Station, reacted strongly to Atkinson's attack on Jerry Franklin and the new forestry scheme, especially after Atkinson published his objections in the Winter 1992 issue of *Western Wildlands*. He wrote George Brown, dean of OSU's College of Forestry, charging that Atkinson's piece was "a vituperative pseudoscientific inflammatory political attack on the Forest Service." Atkinson's remarks were "extreme" and "intolerable" and consistent with statements he had made in the past. Philpot told Brown that Atkinson was damaging both his and the college's reputation. Philpot sent a similar note to Jennifer O'Loughlin, editor of *Western Wildlands*, expressing shock at the publication of Atkinson's article, "Silvicultural Correctness: The Politicization of Forest Science," because of its "factual errors and innuendo obviously intended to excite the reader and bias opinions." He wrote that, while he understood an editor's search for an "array of attitudes toward 'New Forestry,'" he believed Atkinson had exceeded the "bounds of professional propriety."[58]

Dean DeBell, a silviculture forester with the Pacific Northwest Research Station, was another critic, observing that new forestry and ecosystem management were philosophies undergirding forest values other than timber production. Despite institutional and political acceptance of new forestry, many foresters questioned the biological consequences of such management practices. DeBell published his objections to Franklin's proposal in the *Journal of Forestry*, praising the publication for its contributions to silviculture and applauding the journal for advancing the understanding of stand dynamics and for its insistence that silvicultural practices be flexible. New forestry's "naturalistic ideology," however, was spreading erroneous information and forcing management strategies to move toward longer harvest rotations, producing "more appealing landscapes, less slash burning, reduced herbicide use, larger trees, more natural snags, and increased carbon storage." Foresters, DeBell warned, were worried about land being removed from commercial production. More consideration should be given to the social, economic, ecological, and political problems for forest ownerships.[59]

Fred Swanson and Jerry Franklin continued to recommend that ecosystem research should drive forest management practices. They centered their argument in the belief that the extraction of resources was a social as well as an economic "decision within certain biological and physical restraints." In the midst of rapid changes in regional forest policy, there should be no simple prescription for managing complex ecosystems. Managers responsible for the stewardship of public lands should employ what they know about natural systems to shape "sustained-yield management," avoid the use of fertilizer and pesticides, and steer clear of intensive forestry practices. Adopting such strategies would lead to properly designed forest stands, patchwork settings, and healthy stream/riparian environments. To achieve those objectives, Swanson and Franklin urged improved communications between land managers and ecological scientists. The critical issue was developing "social systems for the management of natural resources in an uncertain future."[60]

The northern spotted owl/old-growth controversies that erupted in the 1980s were grounded in physical environments involving multiple ecosystems subject to natural and human disturbances. The consequences of the heavy harvests of old-growth forests on federal land following the Second World War first surfaced in scientific journals in the late 1960s and 1970s when Eric Forsman and others identified declining populations of the northern spotted owl and the bird's affinity for old-growth forests. The relation between the owl and its preferred habitat was linked to similar troubles with anadromous salmon and steelhead and their compromised aquatic environments. The multiple threats to species of all kinds, with ecosystems scientists providing substantive supporting evidence, had become hot-button political issues by the late 1980s, posing dire threats to the conventional notions about resource extraction. Although ecosystem approaches were popular among progressive-minded scientists, industrial interests and their supporters viewed the new environmental science as an economic and political threat to their traditional activities.

Chapter Four
Crises and the Northwest Forest Plan

> *To the logging industry, the protection of the northern spotted owl represents an irrational barrier to economic development. Timber interests have predicted economic doom if the owl is protected. To many environmentalists, on the other hand, the owl symbolizes the struggle to preserve the Earth's dwindling biological resources.*
> —Mark Bonnett and Kurt Zimmerman[1]

Court decisions in the early 1990s related to diminishing stands of old-growth timber and declining numbers of northern spotted owls created a political crisis across the forest districts of Oregon, Washington, and northern California, where judges had placed injunctions against federal timber sales. Andy Kerr, the firebrand conservation director of the Oregon Natural Resources Council, an organization supporting the preservation of old-growth forests, contended that the bird's dependence on old-growth forests created a perfect storm for regional politicians—they "didn't know what to do." Their dilemma worsened when ongoing investigations revealed similar worries about the viability of marbled murrelets and several salmon species. The fallout from those findings preceded the 1992 presidential election season and reverberated in Senate and House hearings in the nation's capital. Politicians from timber states widely criticized the Jack Ward Thomas Interagency Scientific Committee (ISC) report, which was an attempt to resolve the fierce debates over old-growth forests. Despite its congressional opponents, the ISC was the first broadscale plan to conserve species and forests on federal lands in the Pacific Northwest.[2] As this chapter makes clear, the ISC report was only the first in a series of efforts to mediate the differences between industrial and environmental interests over public forest management in the Northwest.

Congress and the Bush administration attempted to move beyond the ISC report when the secretaries of agriculture and interior proposed a five-point plan to protect the owl and "expedite timber sales and avoid the drastic cuts in national forest harvests outlined by the ISC." The secretaries insisted their measure would be in "full compliance" with the Endangered Species Act (ESA) and would achieve a "balance between protection of owl habitat and concern for jobs." As part of the plan, the president would impanel the Endangered Species Committee ("God Squad") in situations where logging plans and the ESA were in conflict. The committee was empowered to permit harvests "without disruption by court challenge." The God Squad would be empowered "to develop a long-term forest management plan for Federal lands." To minimize court-induced job losses to protect the owl, the administration's strategy would provide a temporary solution to continue largely unrestricted logging.[3] The "plan" died aborning without ever being introduced in Congress.

For their part, House Democrats commissioned another study involving four respected scientists, John Gordon, Jerry Franklin, K. Norman Johnson, and Jack Ward Thomas. The group (called the Scientific Panel on Late Successional Ecosystems) was tasked to draft a range of options that would protect old-growth forests and salmon populations and provide predictable levels of timber harvests on federal lands. As members of the panel set about their task, industry opponents wanting full access to public timber dubbed them the "Gang of Four." Using methods similar to investigations before and after, the panel relied on earlier reports and databases, acquired maps of relevant landscapes, and, from the assembled documents, proposed a series of reserves. Their options included a range of provisions, from liberal timber harvests to reserving large habitat areas for the northern spotted owl and marbled murrelet. Among the panel's proposed options were requirements for protecting a network of late-successional forests sufficient for healthy streams and species dependent on old forests.[4]

The Gang of Four's report to Congress was the first to present the trade-offs implicit in creating an ecosystem management plan for late-successional forests. Whichever option politicians chose, it was clear that there would be no "free lunch." Protecting "both individual species and the old forest ecosystem comes at a cost," Charles Meslow argued, and it included considerable reductions in saleable timber. The panel's report provided a "glimpse of what a Pacific Northwest ecosystem management plan might look like and cost." Although the information on old-growth species available to the investigation was limited, the panel's report provided "a credible first approximation of the necessary lands and costs associated with maintaining a functional old forest ecosystem in the Pacific Northwest." Prefacing what would emerge in the future, Meslow thought the history of managing habitat for the northern spotted owl was following an evolutionary path "toward ecosystem management."

Although different in some respects, the Gang of Four assessment fell back on the Interagency Scientific Committee's principles and policies.[5]

In the midst of the public furor over the northern spotted owl and receding stands of old-growth forests, Jerry Franklin published a solo-authored article in *Ecological Applications*, "Preserving Biodiversity: Species, Ecosystems, or Landscapes." His argument, one that would play out in public forums in succeeding months, focused on the failures of saving single species under the Endangered Species Act. Doing so ignored the larger physical environments inhabited by other species and life-forms. The appropriate scales for preserving biodiversity were whole ecosystems, larger landscapes, and entire river basins. Even then, it was impossible to encompass more than a miniscule sample of species diversity. As a participant in the Gang of Four panel, Franklin contended that biodiversity and sustainability required large-landscape approaches as the "*only* way to conserve the overwhelming mass—millions of species." The critical issue was conserving little-known "or unknown habitats and ecological subsystems." Franklin insisted that habitat reserves were critically important to conserving biological diversity, citing the Interagency Scientific Committee report's recommendation for establishing habitat conservation areas for the northern spotted owl. He closed with an appeal "to put all reserves and corridors and heroic mega-fauna in perspective, the larger task of stewardship for all species and all landscapes."[6]

By the mid-1990s, understanding the physical environment was increasingly deemed essential to conservation strategies for fish and wildlife. Fred Swanson and colleagues considered natural disturbances in forest and stream systems in the context of federal laws ranging from the Clean Water Act to the Endangered Species Act, all of which tasked land managers with responsibility for analyzing the effects of human activity on species and ecosystems. With the passing years and accumulated legislation, managers were accountable for environmental conditions in ever-larger spatial scales such as watersheds. When Swanson and his fellow scientists published their findings in 1997, knowledge about Northwest forests had evolved to the point that officials were better equipped to "fit the managed biological landscape with the physical landscape and its historic disturbance processes." The transition to ecosystem approaches to land management would be difficult and would require the public's understanding of the work ahead.[7]

Through this tense period of court injunctions and scientific reports, the Bush administration and Congress were straitjacketed in finding political solutions to the seemingly intractable problem of limiting harvests on federal forests and protecting dependent species. "Nobody, including Congress," the *New York Times* commented in September 1992, "seems to have the stomach for such a task in a political year." Those questions, the *Times* reported, required patience and time that extended beyond the November elections. Democratic presidential candidate Bill Clinton was suggesting federal/business cooperation to accommodate the needs of the environment and to sustain a healthy economy. President Bush, however, wanted to protect the timber industry.[8]

With the presidential election looming on the horizon in the fall, President Bush maneuvered awkwardly through the dueling thicket of court decisions and scientific reports, declaring at the Earth Summit in Rio de Janeiro, Brazil, in 1992 that America's national forests would begin practicing ecosystem management. Back on the ground in the states to campaign for reelection, Bush, referred to "that little furry-feathery guy" and promised to seek adjustments to the Endangered Species Act after the election. That remark prompted candidate Clinton to declare that he would convene a "Forest Summit" to work through the contentious issues and come up with proposals that would pass legal tests and protect communities heavily dependent on federal timber. Bill Clinton won the election in the three-way contest that included Bush and Libertarian candidate Ross Perot.[9]

Living up to his campaign promises, the newly elected president scheduled a "Timber Summit" to be convened in Portland on April 2, 1993. Before the summit, Clinton sent newly appointed Secretary of the Interior Bruce Babbitt on a fact-finding tour through western Oregon. Kathie Durbin wrote in the *Oregonian* that Babbitt "looked, listened, and hiked through the woods," but refused to address the findings that might come out of the Portland meeting. He flew by helicopter to the Andrews headquarters on a late March morning, expressing amazement at the prodigious forestland he passed over. Its rugged productivity, he thought, should provide "room for protecting diverse ecosystems" and for producing enough timber to provide for a stable economy. The secretary visited three places on the Andrews, the log-decomposition site, a high vantage point where he could see a patchwork of different logging practices, and a northern spotted owl nesting area. Before heading for Portland, Babbitt assured his hosts that the Clinton administration wanted to "find a balanced solution" to provide jobs, ensure sufficient timber to keep the mills running, and "do a reasonable job of conserving environmental values."[10]

When President Clinton welcomed attendees on the morning of the Timber Summit, he wanted to find a way through the timber wars that would satisfy the courts. He thanked Oregon governor Barbara Roberts and Portland mayor Vera Katz for hosting the event and Governors Mike Lowry (Washington), Pete Wilson (California), and Cecil Andrus (Idaho) for attending. In addition to Vice President Al Gore, cabinet members Bruce Babbitt (interior), Mike Espey (agriculture), Robert Reich (labor), and Carol Browner (Environmental Protection Agency) were present. They were here, the president said, "to listen and learn." He promised that they wanted to hear voices from all sides, because the issues were important and "intrinsic" to the lives of people in the Pacific Northwest. He wondered, "How can we achieve a balanced and comprehensive policy" that recognized the significance of timber to the region's economy and still preserve the treasured old-growth forests that are an indelible part of the nation's heritage? Clinton closed with the familiar adage, "A healthy economy and a healthy environment are not at odds with each other."[11]

After a long day of testimony, the upshot of the meeting was Clinton's appointment of a Forest Ecosystem Management Assessment Team (FEMAT). Jack Ward Thomas, the wildlife biologist who served as chief of the Forest Service from 1993 to 1996, was team leader of FEMAT; OSU's Charles Meslow headed the Terrestrial Ecology Group; and Jim Sedell was second in command of the Aquatic Ecosystem Assessment panel. The names of Andrews affiliates and Pacific Northwest Research Station personnel are liberally sprinkled throughout the list of participants. Before the year was out, FEMAT, with some six hundred people involved, delivered an enormous thousand-page report offering ten alternatives for managing federal forests. More than any other identifiable group, scientists affiliated with the Andrews Experimental Forest were heavily involved in FEMAT (as well as earlier panels), especially in leadership positions.[12]

The preface to the FEMAT report listed the five principles that President Clinton enumerated to guide their work:

— First, we must never forget the human and economic dimensions of these problems.
— Second, . . . we need to protect the long-term health of our forests.
— Third, our efforts must be . . . scientifically sound.
— Fourth, the plan should produce a predictable and sustainable level of timber sales and nontimber resources that will not degrade or destroy the environment.
— Fifth, to achieve these goals, we will do our best . . . to make the federal government work together for you.

Jim Sedell was an aquatic scientist with the Pacific Northwest Research Station, a member of the Stream Team under the International Biological Program, and later a leader of the Aquatic Ecosystem Assessment Panel of the Forest Ecosystem Assessment Team. Sedell was also an architect of the River Continuum Concept.

The objectives of the president's mandate were to draft management alternatives that would meet legal requirements and achieve "the greatest economic and social contributions from the forests of the region." The assessments were to adopt ecosystem approaches to managing the federal forests in the Pacific Northwest to restore and maintain biological diversity and long-term productivity, including timber harvesting. The report must assure the "maintenance of rural economies and communities."[13] The president's FEMAT directive presented a tall order—fix the seemingly intractable problem of meeting legal environmental obligations and satisfy the needs of the region's timber-dependent population.

In an article in *BioScience*, Jerry Franklin, an important participant in the FEMAT undertaking, characterized the spirit of his colleagues as "Scientists in Wonderland." Those involved, he wrote, had "the satisfaction of working to ensure that decisions are based on the best science available and that decision makers (and society) understand clearly the difficult tradeoffs." Franklin's colleague, Fred Swanson, was equally effusive, recalling that Andrews scientists who specialized in various organisms and topics "were stoked—they got to speak for their taxa (lichens, fungi, mollusks) for the first time in a high-level science-input-to-policy forum."[14] Among the ten alternatives generated in the FEMAT report, President Clinton and his advisers chose "Option 9," which called for the designation of late-successional forest reserves, forest-matrix areas for timber production, and management areas to test various silvicultural practices. The secretaries of agriculture and interior, who signed off on the agreement in April 1994, believed that the principles addressed in Option 9 would protect habitat for the northern spotted owl and the marbled murrelet, the small seagoing bird that nested deep inside old-growth forests.[15]

No one was pleased with the president's choice. The environmentalist Institute for Social Ecology charged that the plan put "a thin veneer of conservation on the continued destruction of the region's forest ecosystems." Timber companies complained that too many acres of federal forests would now be locked up in reserves. Environmentalists, on the other hand, declared that the reserves were inadequate to protect species dependent on old growth for their survival. Option 9, according to the Institute for Ecology, cloaked its questionable ends in progressive language. To gain support for the Clinton plan, the president sent Interior Secretary Bruce Babbitt to the Northwest with promises to immediately release thousands of acres of federal land in timber sales. The carrot for environmentalists, the institute claimed, were vague assurances that they would have a voice in the final plan submitted to the courts.[16]

Jack Ward Thomas explained how Option 9 quickly morphed into the Northwest Forest Plan (NWFP). Because President Clinton wanted to provide a document to the courts that would protect all endangered and threatened species, the plan's buffers were enlarged around seasonal streams, ninety-eight-acre reserves were created around all known owl nests, and the revised policy included a "survey and manage list," protocols for protecting less-known species. For many species, therefore, the revisions to Option 9 required surveying areas targeted for harvesting to determine whether the species were even present. President Clinton's objective was to advance a proposal that would allow Judge Dwyer to lift his injunction on logging in old-growth forests. With those changes to Option 9 (which appeared in the Northwest Forest Plan), the permissible harvests were reduced from 7.3 million board feet per year to 6.4 million board feet.[17]

The NWFP covered twenty-four million acres of federal lands in the range of the northern spotted owl—national parks, US Fish and Wildlife stations, national forests, and BLM lands—in Washington, Oregon, and northern California. The arrangement was designed to protect and restore old-growth forests and aquatic environments to sustain viable numbers of "birds, mammals, fish, amphibians, plants, fungi, and lichens" where habitats had been degraded by decades of logging. In addition to those policy directives, the

proposal included provisions for "predictable and sustainable" federal timber sales. To accomplish those objectives, the twenty-four million acres were
divided into land-use areas (see table below).

Of the land inside the plan, 37 percent was already protected in wilderness
and areas withdrawn for administrative purposes. Under this distribution, 80
percent of existing old-growth forests would be protected in reserves. The
plan also carried a provision for thinning plantation areas within the reserves
to advance the transition to old-growth habitat for the northern spotted owl
and other organisms. Jim Furnish, supervisor on the Siuslaw National Forest,
argues that the NWFP "was huge, complex, and radically different from the
management plans it replaced."[18]

The "survey and manage" measure proved to be the great conundrum for
implementing the provisions of the plan. The measure, added to Option 9
as the plan moved forward, was unprecedented, placing hundreds of lesser-
known species—bryophytes, fungi, lichens, and arthropods—potentially
under scrutiny for federal protection. Survey and manage added rare and
difficult-to-identify species to the list subject to review on forest-matrix lands.
FEMAT had already listed 1,120 species ranging from lichens to arthropods,
fish, and a wide range of animals associated with late-successional and old-
growth forests. Studies focusing on the owl, murrelet, and salmon, however,
did not provide protection for lesser-known species. Because of the lack of
scientific evidence, a 1993 study indicated that areas set aside for the northern
spotted owl provided little security for most of the species in late-successional
and old-growth forests.[19]

One consequence of the Northwest Forest Plan's survey and manage component was the reduction in harvest levels on national forest and BLM forests

Land-use allocations in the Northwest Forest Plan

Land-use allocation	Acres	Percentage
Congressionally reserved areas[1]	7,320,660	30
Late-successional reserves	7,430,800	30
Managed late-successional reserves[2]	102,200	1
Adaptive management areas	1,520,800	6
Administratively withdrawn areas[3]	1,477,100	6
Riparian reserves	2,627,500	11
Matrix lands	3,975,300	16

1 These are wilderness areas, national parks, and other areas set aside before the creation of the plan.
2 Buffers to protect spotted owls and other species.
3 Areas withdrawn from timber sales before adoption of the plan.

from 7.3 million board feet annually to 6.4 million board feet. Under survey and manage, scientists were required to determine whether any (or all) of the species were present and then adjust harvest plans accordingly. The principal landscapes to be considered were forty-five late-successional reserves, or 30 percent of the plan area. Located in proximity to land already protected, the reserves were designed to each host twenty pair of northern spotted owls. Another FEMAT provision carried over to the plan was the Aquatic Conservation Strategy, mandating a series of riparian reserves on both sides of streams, encompassing about 11 percent of the land area.[20]

Survey and manage was complex, time-consuming, and expensive. Its objectives required gathering data about species, locating and protecting their habitats, and developing a plan for protecting them. Initially, specialists conducted field investigations and drafted guidelines for sites identified with specific species. The management recommendations for each species provided information on its natural history and status, and directives for protecting its habitat. The "predisturbance surveys," Randy Molina determined, "were a mitigation measure designed to avoid inadvertent loss of sites that might contribute to species persistence." Surveys focused on 354 rare species across twenty-four million acres of the Northwest Forest Plan. The task, which was to be completed in ten years, was daunting and carried out amid inadequate budgetary resources.[21]

Predisturbance surveys were expensive and controversial. Because some species were identified in thousands of locations, survey administrators lost enthusiasm for continuing such activities. Shutting down surveys was detrimental to timber sales and put the entire program at the center of controversy. And there was still more: when agencies failed to submit timely survey findings, industrial interests sued the agencies, requiring federal officials to conduct environmental impact reviews to extend deadlines. At this point, Northwest Forest Plan officials agreed to revising and redefining the program. Legal challenges continued, with a lawsuit in 2001 forcing another reassessment and analysis of survey and manage and eventually its termination in 2004. Environmental organizations sued, favoring retention of the survey and manage program. By this time, the George W. Bush administration had been in office for a full term, and although the president supported the Northwest Forest Plan, his policies worked on several fronts to remove barriers to timber sales on federal forests.[22]

In its death throes, survey and manage had advanced the scientific understanding of little-known species in the region. After ten years of surveys, field personnel had amassed data on more than one hundred species, with most of the information coming from matrix lands where potential timber sales forced workers to conduct predisturbance surveys. After revisions in 2001, survey and manage focused on reserve lands and late-successional and old-growth habitats. Because many rare species occurred outside the late-successional reserves, the question arose whether the reserves by themselves provided sufficient habitat. Those issues aside, however, scientists concluded that survey and manage advanced knowledge about forest environments: "Whether perceived as a visionary conservation program or simply an experiment of unbridled management complexity, the SM Program accrued important gains in knowledge about rare and little-known species."[23]

As the Northwest Forest Plan was commemorating its first ten years, Jack Ward Thomas and colleagues offered three proposals to move the program toward fully achieving its objectives: "(1) recognize that the NWFP has evolved into an integrative conservation strategy; (2) conserve old-growth trees and forests wherever they are found; and (3) manage NWFP forests as dynamic ecosystems." The third proposition was an attempt to improve the balance between short- and long-term risks. Because the Forest Service and BLM tended to minimize short-term risks, there was great danger of increasing risks in the long term that would lead to failure. All age classes of trees across forest landscapes should be under review, and there should be "species-specific protection" for all endangered and threatened species. Thomas and his team concluded that the dramatic shift in federal forests from sustained timber production to protecting for biodiversity was disruptive to federal agencies and the lives of many people.[24]

Nancy Diaz, who headed BLM's Oregon/Washington office in Portland, called the NWFP an extensive effort to implement a new planning strategy on federal lands, managing whole systems of vegetation, their biophysical environments, and infinitely large numbers of species. What was intrinsically important about the forest plan was its authorship, the broad-based involvement of the scientific community requiring cooperation between scientists and land managers. Diaz reviewed the details of the forest plan, its standards and guidelines, allocations of land, aquatic conservation strategies, monitoring, and adaptive management. Although harvest levels fell far short of estimates, it was too early to call it a failure or a success, because agencies were still struggling with adaptive management, with portions of the plan area subject to experimental strategies. Still, it "was the first plan of its kind to attempt such a comprehensive, ecologically based effort with diverse agencies on a large land base." Diaz pointed to one of the great weaknesses of the plan, the failure of managers to experiment in the adaptive management areas.[25]

After ten years under the NWFP, a team of scientists published an assessment of late-successional and old-growth forests on the twenty-four million acres of northern spotted owl habitat. A satellite mapping of vegetation revealed that assumptions about the extent of older forests were accurate. Despite losses in stand-replacing wildfires, aging forests had expanded and outstripped losses between 1994 and 2003. The scientists' appraisal emphasized that the plan was a strikingly new and innovative approach to managing federal lands. To streamline their assessment, they referred to "older forests" in lieu of late-successional and old-growth forests, according to the Pacific Northwest Research Station report. The abbreviated description offered greater flexibility in assessing degrees of change over time. Because the age of forests was complex and sometimes misleading, scientists relied on measurable attributes, the average size of large trees to determine age.[26]

Using the new definition revealed a net increase in older forests ranging from 1.25 to 1.5 million acres, a figure that exceeded expectations in the NWFP. Defining older forests based on their potential to develop old-growth characteristics could be improved over time as scientists gained a better understanding of older forests. The team underscored the importance of fire in shaping healthy older forest ecosystems. At the scale of the twenty-four million acres, estimates of older forests depended on definitions (using maps and data from plot studies as a baseline for forest conditions). The ten-year survey revealed that 16,000 acres of older forests had been harvested and another 102,000 acres had burned, much of it in the great Biscuit Fire in southwestern Oregon. Given the available data, the authors estimated that the distribution of older forests was greater than anticipated.[27]

A second Forest Service assessment ten years after the adoption of the Northwest Forest Plan focused on monitoring and ongoing research. The report, under principal author Richard Haynes, included a positive note—the area of older forest had expanded to a degree greater than expected, and wildfire losses were within predicted percentages. Although the region's federal lands were expected to "carry most of the weight in conserving species and old-forest ecosystems," Haynes believed nonfederal lands had the potential to assist in meeting the plan's objectives. The aquatic conservation effort, addressing habitat conservation rather than single species, was working well and contributing to improved conditions in many watersheds. Greater numbers of big trees and sharp reductions in clear-cutting were the principal determinants of healthy watersheds. The number of northern spotted owls was still declining, especially in the northern sections of its range, due in part to advances of the aggressive barred owl into its habitat.[28]

Under socioeconomic factors, the Northwest Forest Plan fell short in promising continued federal timber sales *and* protecting old-growth forests: "Timber sale expectations were not met." Old-growth forests were protected, but the services federal forests provided to rural populations in jobs did not meet expectations. Moreover, adaptive management, a critical element in the plan, "proceeded in fits and starts." The promise of experimental silvicultural projects was limited, with regulatory agencies failing to initiate robust and truly innovative projects. Only on the H. J. Andrews Forest and adjacent landscapes were scientists able to arrange collaborative relationships with land-management personnel in a few projects. Although ten years was a limited time for determining success and failure, there were obvious trends in greater protection for old growth, restoring watersheds, and decommissioned roads. Providing a "predictable and sustainable level of timber," however, proved an empty husk.[29]

Expanding the network of late-successional landscapes to enhance biodiversity in plant and animal species was among the wide-ranging expectations of the Northwest Forest Plan. Watersheds linked to the aquatic conservation strategy had fewer roads, which improved the overall health of the environment. Careful monitoring revealed that older forests were naturally developing from early-successional stands, and that thinning experiments accelerated the development of old-growth characteristics. Given their heavy management responsibilities, reduced budgets, and staffing, Forest Service personnel were pressed to the limits. Cutting the number of federal workers in small communities also removed talented and influential people who contributed to the local welfare. Those caveats aside, at the ten-year marker, the NWFP still represented an ambitious long-range effort to responsibly manage federal lands. Haynes cautioned that difficulties loomed ahead: changing administrations, unforeseen economic factors, and "ecological surprises" such as diseases, invasive species, wildfire, drought, and a warming climate.[30]

The purposes of the Northwest Forest Plan were to protect old-growth forests and their dependent species and provide a layer of support for timber-dependent communities. Because the plan was a political compromise, it necessarily embraced both ecological and economic objectives, and because federal forests were vital to local economies, it was important to include community support. Susan Charnley's study of the NWFP provides striking comparative timber-employment figures for the twenty-four million acres before and after the plan was implemented. Before the spotted owl decisions—between 1970 and 1974—timber industry employment provided 36.2 percent of all jobs in northern California, 30.8 percent in southwestern Oregon, and 19.8 percent on the Olympic Peninsula. Fifteen years later (1985–1989), employment had plummeted to 15.3 percent, 13.3 percent, and 10 percent of total employment, respectively. Several issues explain the decline in timber/wood products jobs: economic diversity, mechanization, a shift of timber capital to the Southeast (Georgia, the Carolinas, Louisiana, and Alabama), and the closure of lumber mills in the Northwest during the severe recession of the early 1980s.[31] The important takeaway is that *before* 1990 and the spotted owl decisions, timber industry employment was already in sharp decline.

Under the Northwest Forest Plan, it was assumed that the Forest Service and BLM would resume limited timber sales to offer some support for communities dependent on federal logs. Because reduced timber sales were not expected to return sufficient revenue to county governments—Forest Service sales returned 25 percent and BLM 50 percent—Congress provided offset funds in lieu of declining timber sales. Ten years into the plan, the figures indicate a troubled economy—eleven thousand jobs lost in the plan area between 1994 and 2003, a figure that included both public and private lands. Jobs lost through reductions in federal timber harvests were about four hundred workers. Other jobs reductions, Charnley explains, were due to "mill closures in response to previous supply declines, industry retooling to more efficiently process small-diameter logs, and continued investment in labor-saving technologies." Job losses were also significant in the public sector: the Forest Service lost 36 percent of its full-time workers and BLM 13 percent. With its vast rangelands east of the Cascade Range, BLM was able to transfer personnel to non-timber jobs.[32]

Ten years into the operations of the Northwest Forest Plan, proponents judged it a success, a model for ecosystem management on a grand scale. The effort to address economic and social issues, however, was mixed. The plan's strategy to rely on two-hundred-year-old forests for 50 percent of harvests fell flat in the face of legal appeals and court cases whenever older timber was put on the market. Charnley correctly argues that it is mistaken to believe that a steady flow of logs would provide stability for timber-dependent communities. If the Forest Service and BLM were able to deliver their sales, the supply of raw logs would have been woefully short of sustaining local economies. Federal lands had developed other values as the twentieth century drew to a close, in which citizens trekked to national forests for a variety of recreational purposes, thereby providing alternative economic models for communities adjacent to national forests. A few counties and locales endowed with spectacular outbacks began to flourish through the 1980s and 1990s. As an inclusive design for ecosystem management, Charnley concludes, the NWFP was most successful in the socioeconomic sphere for creating jobs in timber counties related to "restoration, research, monitoring, and other forest stewardship activities."[33]

Jack Ward Thomas, Jerry Franklin, John Gordon, and K. Norman Johnson offered a synopsis of the Northwest Forest Plan in 2006:

> In the 1990s the federal forests in the Pacific Northwest underwent the largest shift in management focus since their creation, from providing a sustained yield of timber to conserving biodiversity, with an emphasis on endangered species. Triggered by a legal challenge to the federal protection strategy for the Northern Spotted Owl, this shift was facilitated by a sequence of science assessments that culminated in the development of the Northwest Forest Plan. The plan, adopted in 1994, called for an extensive system of late-successional and riparian reserves along with some timber harvest on the intervening lands under a set of controls and safeguards. It has proven more successful in stopping actions harmful to conservation of old-growth forests and aquatic systems than in achieving restoration goals and economic and social goals. We make three suggestions that will allow the plan to achieve its goals: (1) recognize that the Northwest Forest plan has evolved into an integrative conservation strategy, (2) conserve old growth trees and forests wherever they occur, and (3) manage federal forests as dynamic ecosystems.[34]

A critical component of the Northwest Forest Plan involved the creation of Adaptive Management Areas (AMAs), an idea prompted by the work of C. S. Holling, Carl Walters, and others beginning in the 1960s and 1970s. Walters was especially interested in "dealing with uncertainty in the management of renewable resources." Holling and Walters were both interested in uncertainty in scientific investigations—in the case of Walters, Pacific salmon

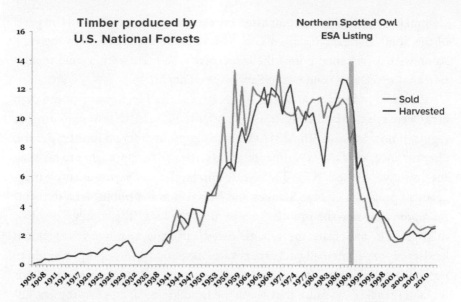

This graph illustrates the sharp decline in timber harvests on the national forest system as a consequence of decisions to protect the habitat of the northern spotted owl. The US Forest Service provided the data reflected in the graph.

management. At the root of adaptive management was the notion to antici-pate the unexpected, to expect surprises, and to accept that "surprising results are legitimate," as Kai Lee argued, "rather than signs of failure, in an experi-mental framework." The ultimate objective of adaptive management was to improve our understanding of management practices through long-term strategies involving what we know about biophysical systems and their inter-play with social and economic systems. Such a process would involve cycles of assessment, implementing experimental treatments, monitoring, further assessments, and revisions until a broader understanding of ecosystems was achieved.[35]

FEMAT posited the creation of ten Adaptive Management Areas encom-passing about 1.5 million acres, approximately 6 percent of the three states in the NWFP. The areas ranged in size from 92,000 to 500,000 acres. Centered in FEMAT's mission statement and consistent with the principles of conserva-tion biology, adaptive management would "encourage the development and testing of technical and social approaches to achieving desired ecological, economic, and other social objectives." The adaptive management landscapes, which stretched from the Olympic Peninsula south through the timbered sec-tions of western Oregon and northern California, included the 325,000-acre Applegate AMA that embraced the watershed of that name. The 158,000-acre

Central Cascades AMA encompassed Forest Service and BLM lands on parts of the South Santiam, Calapooia, and McKenzie watersheds, with the H. J. Andrews Experimental Forest the centerpiece—an area with a solid reputation among scientists and Forest Service personnel.[36]

Adaptive Management Areas and matrix lands (16 percent of the plan area) were the two land classifications under the NWFP that were to provide regular timber harvests. The AMAs were to represent different forest types and circumstances to allow experimenting with alternative approaches to meeting the objectives of the NWFP. Three years into implementing the adaptive management program, George Stankey and Bruce Shindler published a thoughtful report assessing the promise and peril of the idea. The strategy required managers "to anticipate the unanticipated," treating adaptive management landscapes experimentally, to expect surprises, and to respond prudently to experimental findings. The adaptive management initiative posed problems for land managers, because they would be functioning in an environment different from those in which they had operated in the past. Adaptive management would challenge those who were interpreting forestry science in making management decisions. In contrast, the new portfolio included the knowledge of citizens who were familiar with local forested landscapes, views that might directly challenge prevailing forestry practices. Such conflicting opinions provided room for great disagreement over "different ways of knowing," the authors cautioned, although in principle, it offered "the diversity and richness these different forms of knowledge represent."[37]

Stankey and Shindler believed there was no certainty that input from local stakeholders would "result in expanding or challenging our knowledge base." Despite adaptive management's many scientific assumptions and questions, the approach was an abstraction, its premise centered in experience, which supposedly would lead to modifications in the field. Practitioners would be operating "in the face of ongoing uncertainty, with its only reward surprises." Adaptive management would function in a world of multiple viewpoints, where opposing opinions would present constant challenges. Stankey and Shindler questioned whether land managers possessed the breadth of mind to accept the potential of adaptive management. Implementing the program would take time under circumstances requiring immediate results.[38] A skeptic might easily conclude that its premise was pie-in-the-sky optimism.

Despite critics in the Pacific Northwest, there were several adaptive management initiatives under way shortly after the launch of the Northwest Forest Plan. The earliest effort appears in a *Journal of Forestry* article assessing the potential of the Augusta Creek project on the South Fork of the McKenzie River in the Willamette National Forest. The Forest Service's John Cissel and colleagues from the Andrews Forest initiated the study to understand past disturbances and their potential for developing strategies to sustain animal and plant communities into the future. The Augusta Creek area, they believed, was potential habitat for northern spotted owls and other old-growth-dependent species. The objectives of the project were to maintain dynamic landscape patterns within some range of previous historic environmental conditions to sustain native species and ecosystems, and to produce wood fiber. The project would embrace an understanding of historic and present conditions that could support the needs of individual species.[39]

Four years later, Cissel and an expanded team published an extensive study of the Augusta Creek project based on historic fire disturbances. The 19,000-acre planning area was a landscape of steeply dissected valleys, with elevation ranging from 2,300 to 5,700 feet, somewhat comparable to the Andrews Experimental Forest. Tree species in the Augusta area paralleled those on the Andrews—Douglas-fir, western hemlock, and western redcedar, and in drier locations, grand fir, Pacific silver fir, and noble fir prevailed at higher and colder elevations, along with mountain hemlock. The ridgetops at the southern edge of the area were meadows, with conifers beginning to invade the open spaces. Unroaded reserves on the ridges provided markers for the area's northern and southern boundaries.[40]

Focusing on fire regimes during the past five hundred years, the team profiled changes in the Augusta watershed, including human alterations—roads adjacent to riparian zones, clear-cut harvests, the construction of Cougar Dam in the 1960s, and parts set aside in wilderness and roadless areas. The Augusta study included small watershed reserves and valley corridor reserves connecting the watersheds. The project's timber harvests reflected historic fire regimes that would determine the frequency and degree of cutting. The Augusta blueprint projected landscape and watershed management conditions two hundred years into the future. The premise for the project, the landscape's history, was that native species had survived previous disturbances (e.g., fires and landslides), and if those conditions prevailed in the future, it increased the likelihood that native species would survive as well.[41]

The Augusta Creek project was launched amid human-related disturbances that had dramatically altered the South Fork landscape before the Second World War. In addition to Cougar Reservoir, there were hundreds of miles of logging roads and clear-cuts that were now growing even-aged stands

of Douglas-fir. That new environment was not favorable to threatened species like the northern spotted owl and bull trout (*Salvelinus confluentus*). On
the eastern side of the South Fork watershed, portions of the Three Sisters
Wilderness and additional unroaded areas provided a degree of protection
for such species. Because the western side of the South Fork had undergone
significant timber harvesting since the late 1950s, the area was heavily roaded
and covered with second-growth plantations of various ages. The roads on the
western side also contributed to runoff that increased sediment in waterways.[42]

Public land on the South Fork of the McKenzie drainage falls within the
twenty-four million acres of the Northwest Forest Plan and is designated a Tier
1 Key Watershed as primary habitat for bull trout. In the Cissel team's plans,
timber harvests were planned after careful watershed analyses. A late-successional reserve overlapped with the Three Sisters Wilderness in the northeastern
portion of the South Fork watershed, providing habitat for the northern spotted owl. The larger landscape adjacent to the Augusta Creek project, therefore,
served multiple purposes—significant roadless areas, new even-aged plantations, and recent clear-cuts, and a central corridor along the South Fork frequented by campers, fishers and hikers, sightseers, and horse riders. The paved
road through the corridor had been designated a National Scenic Byway, an
attribute attracting motorists and bikers during the summer months.[43]

Cissel and colleagues designed the Augusta Creek/South Fork study
before the adoption of the Northwest Forest Plan. Emphasizing disturbance
regimes, the group hoped to develop management strategies that went
beyond merely understanding past environmental conditions. Although their
approach faced hurdles—it had not been evaluated in a formal environmental impact statement—Cissel hoped the Northwest Forest Plan could be
amended to accommodate their work on the South Fork watershed. Under
adaptive management, all forest plans were experiments of some kind, and
although the Cissel proposal had no precedent, Cissel believed they had
described "an untested assumption that managing within an interpreted range
of historical variability can emulate natural ecosystems and provide a full array
of habitats and ecological processes." The greater Augusta Creek study, the
team contended, offered an innovative way "to improve implementation of
the Northwest Forest Plan."[44]

Among the Northwest Forest Plan's ten Adaptive Management Areas,
the Central Cascades Adaptive Management Area was in a strong position, because it included the prolific research carried out on H. J. Andrews

Experimental Forest and adjacent mountain valleys. Of the Cascade AMAs' 158,000 acres, 15,000 acres was under the jurisdiction of the BLM. Another small fraction, 600 acres, was private and tribal land. Two major watersheds marked the boundaries of the Cascades unit, the South Santiam and McKenzie Rivers. Because the area embraced parts of two large drainage basins, it fit the model for ecosystem management, an appropriate scale for managing natural resources. John Cissel authored a research prospectus for the Cascade AMA: "Intensive research on ecosystem and landscape processes and its application for management in experiments and demonstrations at the stand and watershed level." Its watchwords would be flexibility and providing "room for innovative and experimental approaches."[45]

John Cissel indicated that the Central Cascades AMA enjoyed a head start because of its association with well-established ecosystem research on the Andrews Experimental Forest, two Research Natural Areas, and the successive LTER grants awarded to the Andrews. Cissel wrote an assessment in 1995 pointing to the Central Cascades' potential for research and education, identifying ongoing studies, questions related to ecosystem management, opportunities for future investigations, relations between the Andrews Forest and the Cascades AMA, and illustrating "steps that capitalize on existing research and learning programs." Cissel was writing on behalf of the Cascades Center for Ecosystem Management, a research-management unit formed in 1991, a partnership between the Andrews scientific community and the Willamette National Forest. Although much of the adaptive management area lay outside the experimental forest, an argument could be made that the two places were analogous.[46]

A featured partner in the Cascades AMA was the Blue River Landscape Study, a collaborative effort involving Andrews scientists and Blue River Ranger District staff who were developing a management strategy reflecting natural historic disturbances in the Blue River drainage. Paralleling the Augusta Creek inquiry, John Cissel and Fred Swanson authored an initial review of the Blue River Landscape Study in a 1997 unpublished paper and followed with detailed descriptions of the proposal. The authors were interested in wildfire influences during the previous five hundred years, and timber harvests on the Andrews (1950–1970) and in the rest of the Blue River watershed (1960–1990). The proposal would "test and modify a landscape management approach based on natural disturbance regimes" consistent with the objectives of the Northwest Forest Plan—protecting habitat for species dependent on older forests and associated aquatic systems and providing sales of timber. The management model for the Blue River study would employ "multiple modes of learning," procedures challenging Forest Service personnel to try new and different land-management approaches, to monitor their work, and to change their operations based on observable results. Long-term research on the Andrews Forest would provide helpful insights to understand ecosystems and experiments elsewhere on the Blue River drainage.[47]

The Blue River study was patterned after the Augusta Creek project, where knowledge about historic disturbance regimes provided a basis for planning management on the future landscape. The study included reserved areas and late-successional and aquatic reserves to meet the requirements of the Northwest Forest Plan. Excluding the Andrews Forest, scientists divided the Blue River watershed into areas coinciding with the extent and severity of historic wildfires. Because the frequency and intensity of fires varied with elevation and aspect, those at lower elevations were moderate and caused a modest mortality among stands of trees. Fires at higher elevations were infrequent, but they were intense, killing most trees. The Blue River study, therefore, identified three landscapes: Area 1 (lower elevation where scheduled timber harvest rotations would be 100 years); Area 2 (middle elevations with rotations of 180 years); and Area 3 (high elevations, rotations of 260 years). The watershed would reflect historic wildfire variability, rather than a uniform strategy for all sites. To assess its effectiveness, the design included a complex, long-range monitoring program. Cissel and Swanson termed the study "historically unprecedented" in its approach rooted in landscape dynamics.[48]

The objectives of the Blue River Landscape Study were complex, using historic conditions as a reference point for developing management alternatives. The proposal fit within the geography and mission of the Cascades AMA and its investigations of late-successional reserves to sustain forest ecosystems. Understanding forest history was important to the study, especially the role of historic fire in shaping Blue River landscapes. Fire frequency and severity, and topographic and climatic factors were strategic elements in the plan. High-, moderate-, and low-frequency fires were plotted across the landscape. Based on that data, Cissel, Swanson, and Oregon State University PhD student and fire ecologist Peter Weisberg proposed to overlay silviculture management schemes for off-limits reserves and areas that would be subject to harvest rotations based on historic fires. The authors concluded that management "derived from interpretations of historical disturbance regimes can be carried forward into project implementation."[49]

Like many of the other adaptive management locations, the Blue River Landscape Study and the Cascades AMA were linked to the most intractable and challenging problems under the Northwest Forest Plan. In a perceptive assessment of the AMA system, George Stankey, Bernard Bormann, and Roger Clark edited a collection of articles appraising the initiative during its first decade. The study involved a literature search, fifty interviews with federal land managers and scientists, and more than four hundred surveys circulated to citizen participants to evaluate the viability of the AMAs. "Adaptive management has an appealing simplicity to it," the editors wrote, but "it remains primarily an ideal rather than a demonstrated reality." Multiple barriers prevented its implementation—legal and political requirements to protect threatened and endangered species, the aversion of land managers to taking risks, and institutional constraints preventing people from operating in environments requiring experimentation, learning from investigations, and inviting input from the public.[50]

The ten AMAs were designed to offer wide-ranging biophysical and socioeconomic conditions that would test, validate, and revise management strategies as appropriate. Incorporating citizen input proved difficult, with surveys indicating that less than half of the respondents were aware they could participate. Moreover, less than one-third believed the AMAs were effective in building trust. Despite the agencies' rhetoric that AMA planning would be inclusive, the reality was "business as usual." The editors concluded that agency support for adaptive management was lacking.[51]

The interviews with managers and scientists were revealing—adaptive management lacked clear definition, and there were ambiguities in institutional support for processes such as learning and implementing experiments. Pressure from environmentalists and industry personnel limited the initiatives that agency heads were willing to undertake, and federal agencies themselves mistrusted one another in collaborating on projects. Land managers expressed fear about litigation for violating environmental laws related to endangered species. Stankey and his colleagues reported that interviewees were convinced that regulators, environmentalists, and the public opposed experimental projects for which the results were uncertain. In addition, field personnel frequently mentioned the lack of support. Organizational resources (funding) were less than marginal, opportunities for training were limited, and there was never sufficient time to work on assigned projects.[52]

Editors Stankey, Bormann, and Clark closed their lengthy report on a gloomy note: "There is little on the horizon that signals the agencies are dealing in any substantial way with the barriers discussed in the literature or in this evaluation to cause one to be hopeful." Future adaptive management programs needed to focus on smaller spatial scales to improve the chances for success. Achieving such results would be expensive and require lengthy periods of time, a prerequisite for innovative and experimental programs. The complex biophysical and socioeconomic worlds that adaptive management addresses "require time for effects and consequences to play out."[53]

After a decade operating under the Northwest Forest Plan, it was clear that adaptive management was on the rocks, funding had declined, and it clearly had lost favor among the principal agencies involved. Its diminished stature was disappointing to Andrews personnel, especially to Fred Swanson, who thought the Cascades AMA "was the first high-level, formal assignment to do what we'd been doing for decades." The failures of the program were obvious across the entire AMA system. He cited the initial successes of the Blue River Landscape Study, its logging strategies based on historic fires, the number of visitors to the area, and the observations of social scientists, but those efforts fell for naught in the face of environmental protests, tree-sitters, and lawsuits. It was ironic that both environmentalists and timber companies initiated lawsuits against Blue River field experiments. That said, Swanson observed that basic research programs on the Andrews continued, with new topics being pursued—long-term watershed studies, log decomposition, monitoring of vegetation plots and the northern spotted owl, and a new emphasis on climate change.[54]

With the second decade under the Northwest Forest Plan looming, a team of scientists published a monitoring report "to determine if the NWFP is providing for conservation and management of older forests as anticipated." The appraisal assessed the characteristics of older forests in the plan area and how they had changed since 1993. The scientists developed maps for the beginning and end of the two decades (1993–2012), which they dubbed "bookends." The maps illustrated changes in the extent and distribution of older forests during the study period. The report followed early monitoring documents in which "older forests" were equated with "late-successional and old-growth forests." The twenty-year review addressed the range, spatial arrangements, and changes (disturbances and new growth) since 1995.[55]

The twenty-year evaluation enumerated land-use allocations under the plan: congressionally reserved lands (wilderness areas, national parks, monuments); late-successional reserves (old growth); managed late-successional areas; administratively withdrawn reserves (scenic and backcountry); adaptive management areas (experimental manipulation); and matrix lands (where most timber harvesting would take place). Riparian reserves, another land-use category, were not included because their placement was to be adjusted after further analysis of specific sites. Field measurements of inventory plots in the spotted owl study area would determine the characteristics of forest land. The monitoring program indicated that analyses of older forests revealed a slight decline (2.8 to 2.9 percent) on federal lands, with wildfire the principal cause. The review found slight differences between the map and plot-based studies of decline in older forests, the map estimates being a bit more negative. Although losses were negative, there were areas where characteristics of older forests were beginning to emerge, thereby indicating net gains in areas of older forests.[56]

Connections among older forest stands, important to improving habitat for endangered species, were "further below the outcomes set forth in the Plan," and recent large fires had degraded those connections. The Northwest Forest Plan anticipated wildfire losses in older forests, but their increasing frequency and severity in fire-prone landscapes was a matter of great concern. Although the twenty-year assessment revealed modest changes in older forests, two conditions could significantly accelerate or slow their modification—large forested areas developing the characteristics of older forests and an increase in forest fires. Other issues could also pose problems for future monitoring programs. Although advanced technology made possible increasingly accurate and efficient monitoring, federal budget reductions were already limiting measuring work in the field. Jim Pena, in charge of Pacific Northwest Region 6, put a positive spin on the effort: "Twenty years of monitoring demonstrates our commitment to adaptive management and fulfilling our commitment to the plan itself."[57]

With environmentalists, industrialists, and a core group of scientists questioning the value and purpose of characterizing lands as ecosystems, it is important to examine the meanings that scholars have attached to the words. The adaptive management plans reviewed in this chapter reveal the tensions between ecosystem management proposals and critics from the political right and left. Because the term ecosystem has been politicized, its definition is in the eye of the beholder. That said, healthy ecosystems usually exhibit considerable biodiversity, illustrating the importance of the intersections among vertebrates, invertebrates, and the environments they inhabit. Maintaining biological diversity, in turn, requires that humans manage themselves appropriately in relation to the natural world. Managing ecosystems mandates that we focus on large-scale, holistic environments. Scientists in the Andrews Forest, for example, moved beyond the single focus on fiber production in Douglas-fir to belowground root life and its contributions to aboveground ecosystems. The awareness and appreciation for this new approach to science began with Arthur Tansley in 1935 and haltingly pressed forward in the succeeding decades, confronting detractors at every step of the way.

Chapter Five
Ecosystem Management Under Fire

> *The Clinton administration is merrily making matters worse. By putting most remaining Northwest forests in preserves, to save owls and "ancient" trees, it will eventually destroy the very landscape it seeks to save. This is the Rice Crispies [sic] of ecological salvation—a seemingly sweet idea for making forests and towns go "snap, crackle and pop."*
>
> —Alston Chase[1]

The northern spotted owl would be an important test for ecosystem management, because it was one among many species involved in environmental controversies at the close of the twentieth century. Sensitive, threatened, or endangered animals—gray wolves, grizzly bears, North American lynx, Pacific fishers, greater sage grouse, marbled murrelets, brown pelicans, several salmon species, bull trout, rockfishes, killer whales—and innumerable plants were a few of the species in question. What had slowly emerged since the 1960s and 1970s was the association between vulnerable species and their habitats, the physical environments necessary for their survival. In the telling of that story, the most notable species was the northern spotted owl. When the US Fish and Wildlife Service declared the owl threatened in 1990, it associated its vulnerable status with diminishing habitats—old-growth forests in the Pacific Northwest. Those findings represented one step in the shift of ecological studies from single species to the complexities of larger ecosystems, a recognition that prompted ecologists to better understand the complexities of physical, biological, and social structures.[2]

A parallel development, the mounting environmental accounts of sullied global landscapes and waterways during the 1970s, brought strong reactions from conservative intellectuals. Among the most strident voices was University of Illinois economist Julian Simon, who argued that negative stories about population growth, compromised natural resources, and degraded environments were "based on flimsy evidence or no evidence at all." The truth was just the opposite, Simon argued, as he launched a full-blown assault on Paul Ehrlich's best-selling book, *The Population Bomb*. He charged that per capita food production in developing countries was increasing and that the World Bank's claim that population growth would lead to declining economic development was mistaken. Empirical studies, Simon contended, found "no correlation between a country's population increases and per capita growth."[3]

Developing economies, where people brought "mouths and hands into the world but also heads and brains," provided the substance for much of Simon's argument. The human mind, he believed, contributed to greater knowledge and technical advances. A proper reckoning with economic theory would show that having more children was a positive influence on the standard of living. Simon also warred with the notion that urban expansion in the United States was rapidly eroding the nation's valuable cropland. He cited United States Department of Agriculture figures indicating that draining swamps and reclaiming arid lands far surpassed acreages lost to urban growth. Who was responsible, then, for pushing this bad, scary news? The fault rested with agencies that funded scholarly studies and the market's propensity for bad news to sell "books, newspapers, and magazines."[4]

While Julian Simon was the conservative's poster boy, ecologists in the Pacific Northwest, struggling against habitat loss and ever-larger clear-cuts on national forests, linked troubled habitats to uninhibited industrial activity. To counter threats to species loss and risks to biodiversity, scientists, many of them affiliated with the Forest Service, proposed ecosystem approaches to protect against further erosion of critical landscapes. Lumber industrialists countered that ecosystem approaches meant locking up valuable forest lands in reserves. For their part, most scientists viewed ecosystems as physical/biological environments for investigating natural worlds through time and space. Alston Chase, one of the foremost critics of ecosystem science, associated the idea with nature worship and biocentrism, a philosophy that he claimed gave license to proponents to build "a new misanthropic ethic of nature to replace the older, humanistic one."[5]

Holistic ecosystem approaches took root in the northern West with two groups of totems, the grizzly bear and wolves in the Northern Rockies, and the northern spotted owl in the great Douglas-fir forests of the Pacific Northwest. The association of the owl with old-growth forests, while focusing on the needs of a single species, opened the eyes of scientists to broader relationships among all natural elements in a given environment, or "ecosystem." Coined by the English botanist Arthur Tansley in 1935, ecosystem "referred to a holistic and integrative ecological concept that combined living organisms and the physical environment into a system." Tansley's concept evolved over the years, eventually finding a home in the International Biological Program (IBP) in the late 1960s and early 1970s. Designed to promote the worldwide study of representative samples of the Earth's ecosystems, the IBP's terrestrial communities section eventually centered on biome studies in the United States, including the coniferous forest biome investigations in Oregon and Washington funded under the auspices of the National Science Foundation (NSF).[6]

The purpose of the Coniferous Forest Biome Project, University of Washington's Robert Edmunds reported, was to develop ecosystem theory to define "the structure, function, and behavior of natural and manipulated coniferous forest and associated aquatic (stream and lake) ecosystems." Jerry Franklin argued that it changed the trajectory of research on the H. J. Andrews Experimental Forest: "It was the predecessor of all that went afterwards, because everything that came afterwards was clearly derivative of what happened in the IBP." The Coniferous Forest Biome Project introduced scientific approaches that were new to Andrews scientists. "I didn't know about ecosystem science," Franklin admitted, "but it sounded neat." Ted Dyrness was equally candid: "We were just starting off from ground zero as far as this emphasis on ecosystem, multi-disciplinary research." Although his colleagues "didn't know much about those fields," Dyrness told an interviewer, they wanted "to get on board." To inform themselves, Franklin, Dyrness, and others attended a short course on ecosystem ecology at the University of Wisconsin.[7]

For Andrews Forest personnel, participation in the Coniferous Forest Biome Project marked the beginning of ecosystem science inquiries. In the midst of IBP-funded research, Jerry Franklin spent two critical years in Washington, DC, as director of ecosystem science with the National Science Foundation, where he was instrumental in rolling over IBP funding into continuing support for long-term ecosystem research. Franklin helped guide the IBP initiative into continuing ecosystems research at the Andrews, the Coweeta Hydrologic Laboratory in North Carolina, the Hubbard Brook Experimental Forest in New Hampshire, and other sites. NSF's David Coleman contended that the lasting influence of the IBP on ecosystem ecology is reflected in the long-term ecological research (LTER) investigations that have been carried on at several sites well into the twenty-first century.[8]

Although ecosystem analysis had limited influence in national forest management through the 1970s, that began to change in the 1980s when the habitat requirements of the northern spotted owl became better known. In a keynote address celebrating fifty years of research at the Coweeta Hydrologic Laboratory, Jerry Franklin offered his personal observations on ecosystem research, including the astounding notion that 20 percent of forest biomass was underground. Equally important findings were the significance of forest canopies to transpiration, and the ways that fog drip and cloud moisture contributed to the health of forest ecosystems. Canopies also played a role in absorbing atmospheric pollutants. Franklin reviewed investigations at Coweeta and the Andrews, where research involving forest/stream interactions discovered the important contributions of vegetation and woody debris to healthy aquatic environments. Experiments at Hubbard Brook demonstrated the larger influence of vegetative growth following catastrophic disturbances in preserving nutrients in the soil.[9]

The larger significance of Franklin's presentation focused on developments that advanced ecosystem research. The long-term databases at Coweeta and the Andrews Forest, and the interdisciplinary, holistic research pioneered at Hubbard Brook, made ecosystem science respectable. Franklin was critical of ecosystem theorists, musing that their insights lacked broader significance in the real world. Scientists erred when they disparaged natural history approaches, because they treated the natural world as deterministic. "I view the world as having a lot of randomness," Franklin argued, and scientists needed "to accommodate the great diversity inherent in the world's ecosystems." Ecological theory should be robust and flexible, encompassing much larger temporal and spatial scales. Comparative analysis, the collaboration of interdisciplinary scientists, and the maintenance of long-term databases were critical elements to ecosystem inquiries. Franklin issued an appeal for granting agencies and scientists to protect and maintain existing research sites.[10]

Fred Swanson and colleagues expanded on Franklin's insights, pointing to agents that were disruptive to ecosystems: the long-term consequences of intensive timber management, the conservation of single species, and the looming threat of climate change. Uncertainties abounded about the accumulative effects of present forest management practices. How would evolving public values influence the management of natural resources? And global temperatures? Swanson and colleagues addressed strategies to cope with disruptions to ecosystems, citing the need for large-scale planning, implementing adaptive management practices, and improving their knowledge of natural systems. Scientists should develop holistic strategies to cope with problems in the region's national forests.[11]

The Rocky Mountain Forest and Range Experiment Station joined the ecosystem conversation, chartering a team of management and research biologists to draft protocols for ecosystem management. Their report recommended blending human needs and environmental values to produce "diverse, healthy, productive and sustainable ecosystems." The Rocky Mountain Station's assessment was self-critical, observing that federal scientists should look beyond short-term approaches to environmental problems, avoid favoring single wildlife species at the expense of others, and move "purposefully toward a more holistic form of managing ecosystems for long term sustainability." Human injury to ecosystems, the station's team acknowledged, was a consequence of pursuing short-term profits, practices that should shift toward sustaining viable ecological and economic activities. The group's "Guiding Principles" declared that humans are an integral part of ecosystems, that all abiotic and biotic elements should be protected, that large landscapes would ensure diversity, that commodity production and resilient ecosystems should be maintained, and that management "should conserve or restore natural ecosystem disturbance patterns."[12]

David Bengston of the North Central Forest Experiment Station in St. Paul, Minnesota, joined the discussions about forestry, ecosystem analysis, and spotted owl court decisions. The controversies, in his view, reflected values such as economic growth, faith in technology, and the abundance of resources, ideas juxtaposed against concerns about sustainability, the finite nature of resources, and skepticism about technological fixes. Older values about sustained-yield forestry, multiple-use laws, and the primacy of timber production on federal forests, Bengston contended, had been challenged during the 1980s. Foresters, who were preeminent among the critics, understood "the dynamics of forest ecosystems [and] raised questions about the impacts and sustainability of forest practices." Traditional commodity beliefs in forestry and range management in the Forest Service had given way to practices that forwarded "the protection of ecosystem health, such as ecology, wildlife, biology, and soil science."[13]

One of the earliest book-length studies of ecosystem science, R. Edward Grumbine's *Ghost Bears: Exploring the Biodiversity Crisis* (1992), illustrated the confused meanings of the term. Following the policies of Gifford Pinchot, Forest Service managers viewed ecosystem approaches as a threat to "multiple-use commodity production." Scientists, who were divided over its implications, thought it meant "the processes of nature unencumbered by humans or a model for biodiversity that explained the dynamic processes of life on earth." From those divergent meanings, scientists moved to protect biodiversity from the threats of commodity producers. Referring to Jerry Franklin's *Ecological Characteristics of Old-Growth Douglas-Fir Forests*, Grumbine observed that understanding biodiversity required "thinking like a mountain," considering "biotic elements of plants, animals, and other living things [and] . . . patterns and processes that shape volcanoes and forests."[14]

What we understand about ecosystems should not remain fixed, Grumbine argued, because ecological theory itself was being transformed in a natural world subject to continual change. Today's plant and animal communities were evolving, a reminder that global temperatures have never been in a steady state. The author quoted historian Donald Worster, who addressed the erratic nature of the world, "discontinuous and unpredictable" and full of random events. In the midst of accelerating climate change and rising temperatures, Grumbine predicted that plants and animals would be migrating to higher elevations for more favorable conditions. "In an untidy world, humans can no longer expect uniform solutions to complex problems." Given those projections, Grumbine urged protecting biodiversity, slowing population growth, and challenging "the ethic of domination." If humans failed to understand the world in which they have appropriated everything to themselves, "other species and ecological processes" would be degraded.[15] In the nearly thirty years since the publication of *Ghost Bears*, R. Edward Grumbine's prescription for changing human behavior has largely been ignored.

The Society of American Foresters' venerable *Journal of Forestry* devoted two special issues to biodiversity and sustainability in 1993 in the midst of the forest wars in the Pacific Northwest. Jerry Franklin, with the University of Washington's College of Forest Resources since 1986, led the discussions, urging foresters to see ecological value in trees beyond their age of "economic maturation." Just as human development did not end with adolescence, land managers should take the long view and "integrate ecosystems and landscape science in field applications." Bruce Lippke and Chadwick Oliver, both from the University of Washington, warned about the economic perils of managing for biodiversity. Such regulatory mandates would bring job losses and wreak environmental harm. Landowners would most likely cut their commercial timber in lieu of managing for biodiversity, because chance encounters with endangered species would prevent the owner from harvesting the trees.[16]

Responding to questions from the *Journal of Forestry*, James Lyons, President Bill Clinton's assistant secretary of agriculture, argued that managers of the national forests in the Northwest had embraced sustainability. The Forest Service's new management strategies had taken into consideration "the value of ecosystems and seeks to maintain the health and productivity of ecosystems and the resources that comprise them." Innovative forestry programs depended on presidential leadership and administrations working with Congress to fund such programs. The Pacific Northwest Research Station's Dean DeBell and Robert Curtis disagreed with Lyons, warning readers that ecosystem management and biodiversity would hamper timber production. They questioned the biological principles and consequences of implementing "new forestry" on public lands, urging managers to be wary about "untested practices." When they implemented management decisions, foresters needed to reckon with economics and the guiding principles of social policies.[17]

Three scientists affiliated with the Oregon State University campus—William McComb and William Emmingham of the Forest Science Department, and Thomas Spies with the Pacific Northwest Research Station—addressed proposals for integrating mature-forest conditions into timber harvesting practices. They offered different silvicultural approaches for integrating wildlife habitat on four different stands of timber—single storied (a modified clear-cut), few storied (two or three age classes), many storied (developing old-growth characteristics), and mature forests (restoration of old growth). The authors dubbed their management strategies "landscapes by design," policies to develop forests on ownerships with different types of human disturbances. Their design would minimize fragmenting habitats and encourage connecting mature timber stands. Such landscapes would provide a mosaic of trees at different stages of growth, with the ultimate goal of achieving a mature forest. Harvesting would increase when landscapes were fully managed. The authors anticipated that harvests would depend on the quality of sites, the species involved, and the "relative value placed on wildlife by landowners, and legislative mandates."[18]

Members of the Society of American Foresters, who expressed reservations about the Forest Service's new direction, represented a broad range of professionals—private industry and lumber trade organizations, state and local public foresters, and those with the Bureau of Land Management and Forest Service. Although the *Journal* usually spoke for foresters associated with production-oriented silviculture following the Second World War, that began to change with passage of the National Environmental Policy Act (1970), the Endangered Species Act (1973), the National Forest Management Act (1976), and other environmental legislation. The spotted owl/old-growth controversies of the late 1980s and subsequent court decisions prompted the Forest Service to shift from exclusive production toward management practices that paid more attention to natural forest environments.

A *Journal of Forestry* special issue in 1996 focused specifically on ecosystem management. Jack Ward Thomas, then chief of the Forest Service, and Susan Huke of the agency's ecosystems management staff, cited their understanding of ecosystem management: "A concept of natural resource management wherein national forest activities are considered within the context of economic, ecological, and social interactions within a defined area or region over both short and long term." The Forest Service program for the national forest system would involve ecosystem management, a policy that would meet human needs and protect the health and productivity of federal landscapes. Because ecosystem management was primarily about humans, it was important to involve the public.[19]

Thomas and Huke cited the Blue Mountains in Oregon and Washington as an example of areas where excessive logging and overly aggressive fire suppression had shifted forest ecosystems "away from their historical ranges of variability," natural reference points for characterizing the dynamism of ecosystems. When ecosystems were protected from disturbances (through fire protection), landscapes were prone to adjust through catastrophic fire. Other problems involved ecosystems across multiple land ownerships, conditions that necessitated region-wide planning and the need for managers to work across administrative boundaries. The Forest Service, Thomas and Huke indicated, was actively pursuing such policies with state, private, and tribal jurisdictions. The authors urged land managers to embrace sound, dependable science to frame their management plans—"how ecosystems function; how they can support and tolerate human use; and how policies and management decisions affect resource use, the environment, and recovery."[20]

Because ecosystem management was "a 'fuzzy' questionable concept," Thomas More, a social scientist with the Northeastern Forest Experiment Station, cited the need for a precise definition. "Scientific management proceeds from principle to application," and because ecosystem management lacked scientific principles, it was unscientific. Thomas More worried that administrative agencies, free to interpret the concept as they saw fit, would leave the terms sustainability and healthy ecosystems as vacuous and without meaning. Administrators who used ambiguous and confusing concepts were responsible for recent lawsuits regarding forest management.[21]

Roger Sedjo, a scholar at Resources for the Future, was another critic who thought ecosystem management had merit as a philosophical principle but was limited in "providing operational directives." Sedjo appealed for a return to multiple-use, sustained-yield management strategies of the past: while the Multiple-Use Sustained-Yield Act of 1960 offered a full range of benefits, including sustained-yield timber harvests, ecosystem management provided only a single purpose, forest health. Sedjo accused proponents of ecosystem management of treating Native Americans as part of early ecosystems, but not accepting "the impact of modern humans." The Forest Service's embrace of ecosystem strategies was a reaction to court decisions—the Endangered Species Act, National Environmental Policy Act, and other environmental legislation. Current policy directions, Sedjo feared, were fraught with danger for an agency that relied on the timber industry for its budget.[22]

Thomas More and Roger Sedjo's opposition to ecosystem strategies reflected the view of a cadre of professional foresters who supported traditional approaches favoring producers of wood fiber. Neither author reckoned with legislation that had contributed to the crises of the 1980s and 1990s. They ignored the proposals of Jerry Franklin and others who articulated ecosystem management strategies, albeit under different silvicultural practices. Sedjo thought there was little public support for current Forest Service policies and that multiple-use laws had not been superseded. Management objectives, he insisted, should provide "goods and services to the nation's citizens and taxpayers."[23]

Alston Chase joined the ecosystem science debate in 1995 with a full-blown critique of the concept and its proponents. Author of the controversial *Playing God in Yellowstone*, his new book, *In a Dark Wood: The Fight over Forests and the Rising Tyranny of Ecology*, opened with a review of the cultural ethos that emerged in the 1960s:

> An ancient and philosophical theory, ecosystem ecology, masquerades as a modern scientific theory. Embraced by a generation of college students during the campus revolutions of the 1960s, it had become a cultural icon by the 1980s. Today, not only

does it infuse all environmental law and policy, but its influence is also quietly changing the very character of government. Yet, as I shall show, it is false, and its implementation has been a calamity for nature and society.

Ecosystem science was the villain in an environmental scenario that the public did not understand, Chase contended, fracturing debates over preservation and incorporating new and confusing belief systems. The real problem, according to Chase, was an ill-informed public.[24]

In a Dark Wood deserves attention, because its 535 pages are exhaustively researched; the writing is erudite, stylistically clear, and succinct; and it carries the imprint of one of the nation's most prestigious publishers. Chase expresses his gratitude to a host of people associated with both sides of the ecosystem debate, including scientists with varied affiliations with the Andrews Experimental Forest (Art McKee, Eric Forsman, Jerry Franklin, Charles Meslow, Norm Johnson, and Tom Spies). He thanks other prominent individuals; environmental activists Judi Bari, Dave Foreman, Andy Kerr, Andy Stahl, and Jeff DeBonis; and a host of industry representatives and organizations. Among the latter are the Northwest Forest Council and the Oregon Lands Coalition. His "Prologue: *Dramatis Personae* in Chronological Order," provides brief sketches of some of the personnel featured in the book. The first chapter, "The Search for Nature," begins with Rachel Carson's *Silent Spring*, the growing list of creatures facing extinction, and federal regulations that added "staggering costs to the economies of some regions." Instead of offering citizens hope for the future, Chase accused the environmental movement, responsible for the restrictive legislation, for having "lost its way." *In a Dark Wood* explains "what went wrong."[25]

Chase links his argument to the cultural difficulties of the 1960s and students who rejected the values and politics that had served the nation since its founding. Resorting to direct action, young people rallied around new representations of nature. "These pilgrims discovered ecology," he contends, "based on the notion that nature was organized into networks of interconnected parts called ecosystems." Blending what they knew about ecosystems, the radicals forged a biocentric ideology that valued the health of all living things, granting ecosystems precedence over their constituent parts, including the welfare of humans. The biocentrists, according to Chase, believed society should be restructured to meet the requirements of the new model. By the 1980s, diminishing forests, threatened and endangered species, and other problems with ecosystem components offered a route to reform. Forests, "nature's seemingly perfect ecosystems," he argued, were the most compelling place to bring about change.[26]

The most famous of the struggles "erupted over the 'ancient forests' of the Pacific Northwest and their now famous occupant, the northern spotted owl." The conflict centered on Douglas-fir forests in Oregon and Washington, and the great redwoods in northern California. Chase traced the convulsive clashes that rocked the region and reverberated all the way to Congress. In the fight to save northern California's redwoods, the battle extended to Wall Street. At its epicenter, the struggle involved conflicting ideas about nature and the health of ecosystems, "an ambiguous word that no one understood." As the years passed, ecosystem ecology gained political influence that would end "in tragedy for society and nature."[27] To flesh out this gloomy scenario, *In a Dark Wood* features numerous references to Jerry Franklin, Jack Ward Thomas, and Eric Forsman, three scientists who had contributed to ecosystem ecology.

If there is substance to Alston Chase's thesis of a straight line between 1960s college radicals and the emergence of ecosystems ecology in the 1990s, the backgrounds of Eric Forsman and Jerry Franklin provide striking counterpoints. Son of a Eugene carpenter and a mother who was an enthusiastic birder, Forsman grew up fishing and hunting and, following his mother's passion, developed an affection for raptors, especially owls. He enrolled in the fisheries and wildlife program at Oregon State University, graduating in 1970, and then completed a two-year stint in the army. While serving summers on seasonal fire crews in the Willamette National Forest, Forsman discovered and became adept at calling spotted owls within close range. Returning to school in 1972, he spent much of his time in the field identifying locations of spotted owls, virtually all of them in large, old-growth Douglas-fir forests. When he discovered that owl populations were declining, he attributed the cause to diminishing old-growth timber stands. With two graduate degrees in hand and a handful of articles in scientific journals, Forsman would spend the remainder of his professional career as the principal expert on the northern spotted owl.[28]

Chase draws heavily on the work of Jerry Franklin, who spent his growing-up years in Camas, Washington, where his father worked in the big Crown Zellerbach paper mill. Franklin spent a lot of time in the woods as a youth, especially in the nearby Gifford Pinchot National Forest, where he developed an interest in forestry. He eventually earned bachelor's and master's degrees in forest management at Oregon State University, and a PhD in the College of Forestry at Washington State University. He spent time during the summer of

1959 on the Andrews and developed a great fondness for the place as a setting for a wide array of investigations—plot studies, forest/stream interactions, woody debris on the forest floor and in waterways, and eventually old-growth forests. Nothing in the background of Forsman, Franklin, Fred Swanson, or any of the contributing authors to the important book by Franklin et al., *Ecological Characteristics of Old-Growth Douglas-Fir Forests*, suggested links to radical youth insurgencies in the 1960s. In the cases of Forsman and Franklin there was, however, an ingrained familial fondness for the outdoors.[29]

In a Dark Wood strangely misreads the seminal publication, *Ecological Characteristics*, reproaching the authors for writing a "curiously static" text and tilting their narrative toward discredited ideas about the balance of nature. Oregon State and Pacific Northwest Research Station scholars, Chase wrote, pay "little attention to the history of disturbances in the region." That charge is striking, because at least since the great Christmas flood of 1964, Andrews' scientists had been investigating the history of disturbances on the experimental forest. Ted Dyrness published an article on disturbances associated with the 1964 flood, bringing attention to evidence of similar events earlier in time. Fred Swanson established his scientific credentials on the Andrews and elsewhere delving into historical disturbances in the distant past. Jim Sedell, with the Stream Team on the Andrews and one of the architects of the River Continuum Concept, was another scientist who understood the disturbance history of streams. More important, and what Chase ignored, was that the purpose of *Ecological Characteristics* was to describe existing stands, not landscape disturbances.[30]

Alston Chase was especially appalled at the litigation taking place in the late 1980s and early 1990s. When the Forest Service or Bureau of Land Management advertised a timber sale, environmental organizations would seek legal redress in court. Chase counted more than twenty lawsuits between May 1990 and October 1991, citing the Sierra Club Legal Defense Fund in Seattle and the Environmental Law Clinic at the University of Oregon as principal litigators. The most significant case, *Portland Audubon v. Hodel*, filed in 1987, eventually morphed into *Northern Spotted Owl v. Hodel* in 1990, a suit that asked the Seattle District Court to set aside old-growth habitat for the northern spotted owl on BLM timberland. The agency complied the following year, placing fifty-two timber sales off-limits to harvesting. In a parallel case directed against the Forest Service, *Seattle Audubon v. Robertson*, federal District Judge William Dwyer in Seattle placed an injunction prohibiting timber sales on seventeen national forests in Oregon, Washington, and northern

California until the agency submitted an environmental impact statement protecting the owl. In subsequent lawsuits against federal agencies, most of them successful, the litigants received compensation, because they were intervening on behalf of the "public interest."[31]

The lawsuits were controversial, because scientists outside the Forest Service were challenging the agency's authority, forcing it to confront promises of timber sales they could not deliver. Ecology's new modus operandi forced Forest Service officials away from timber production to emphasize participatory and collaborative approaches to problems. For natural resource agencies, especially the Forest Service, the adoption of biodiversity and minimum viable populations (vested in the National Forest Management Act) challenged traditional management strategies. Environmentalists relied on three legal statutes to address biodiversity: the Endangered Species Act, the National Environmental Policy Act, and the National Forest Management Act. Environmental organizations were successful in taking the mandates in these laws to friendly federal courts.[32]

There were other public policies that offended Chase. The Clinton administration—with second-in-command Vice President Al Gore leading a "reinventing government" initiative—adopted biocentrism as a fundamental principle to guide federal land-management practices. Chase criticized President Clinton for signing the International Convention on Biological Diversity in June 1992, committing the United States to protect "ecosystems, natural habitats and the maintenance of viable populations of species in natural surroundings." In a colossal exaggeration, Chase contended "the treaty set off a tidal wave of planning to analyze and control every square inch of American real estate," setting in motion mechanisms to catalog everything in the natural world. He fails to point out that the Senate never ratified the treaty.[33]

In its closing chapter, In a Dark Wood indicts the environmental movement for being "hijacked by ideas and values" that destroyed the objectives it sought. There were too many ambiguities in the environmental crusade, including its infatuation with wilderness and a disposition to restore an early American world that never existed. Furthermore, according to Chase, environmentalists borrowed from European philosophies and transformed "ecology from a promising science into a highly political one." Returning to his original thesis, he castigates environmentalists for adopting a new doctrine, biocentrism, emphasizing the health of nature and ignoring the welfare of humans. Ecosystems became the prototypical objective, with global survival dependent "on promoting 'diversity' by social engineering or by force."

The author accuses "the movement" and its science for being wrong on key points: "America is not running out of trees; old growth still covers most of its historic range; owls are probably not disappearing and may not even need old growth."[34]

Few arguments *In a Dark Wood* are in concert with the many decades of research taking place on the H. J. Andrews Experimental Forest. Reading the book takes one into a world far removed from the decades of research on Lookout Creek. Art McKee, Andrews site coordinator during Chase's visits to the forest, treated the writer generously, despite warnings from National Park Service personnel that he was not to be trusted. McKee took him to research sites and shared information about ongoing investigations. He visited Chase twice at his home in Livingston, Montana, to review chapters and suggest revisions, "none of which were ever in the final revision of *In a Dark Wood*." Bill Ferrell, on OSU's College of Forestry faculty, also accompanied Chase on field trips and, according to McKee, was outraged by what appeared in print. McKee surmises that Chadwick Oliver—then with the College of Forest Resources at the University of Washington—convinced Chase that Andrews personnel "suffered from a 'cult of personality,' [and] were led astray by Jerry Franklin," who had a tendency "for presenting plausible hypothesis as 'proven,' something we often accused him of doing."[35] Whatever the author's purpose, *In a Dark Wood* painted a grim description of ecosystem science and the Andrews scientists who were prominent in implementing ecosystem principles in management practices.

In the midst of Chase's storm of criticism, Andrews scientists turned their attention to a major flood event in early February 1996, when a subtropical storm accompanied by rising temperatures settled over northwestern Oregon for four days (February 6–9), dumping record amounts of rain. With freezing temperatures rising to 7,000–8,000 feet elevation, the combined forces of rain and melting snow turned mountain streams into raging torrents, driving downstream rivers to flood stage. Gauges on the McKenzie River at Vida, indicating a flow of 4,000 cfs (cubic feet per second) on February 5, increased to 20,000 cfs the next day and peaked at 25,800 cfs on February 9. The Willamette and Columbia Rivers both reached near-record flood levels. The volume in most western Oregon streams did not reach the peak flows of December 1964, although the Tualatin, Clackamas, and Mohawk Rivers set records at some stations in 1996. H. J. Andrews scientists estimated that the

1996 storm was "a larger event at low elevations and a smaller event at higher elevations than in 1964."[36]

For disturbance scientists, the 1996 flood was an exciting event. Looking at his computer screen in Corvallis, Gordon Grant noticed that a monitoring station high in the Lookout drainage, measuring the heavy, wet snowpack, indicated that it held an immense amount of water. (The snowpack was 112 percent of average for that time of the year.) Alerting his colleague, Fred Swanson, about the rapidly rising flow at a gauging station on the McKenzie, the two piled into a four-wheel-drive pickup with three colleagues and headed for the Andrews Forest, where they spent the next two days. The heavy rain darkened the drive above the town of Blue River, where a side road led to the Andrews Forest. At first they noticed small debris flows and water draining from a hillside and across the road. The streams, however, were high and turbulent, and when the men stopped at a bridge over Lookout Creek, they heard the roar of the stream and the booming of huge boulders bouncing along the bottom. With darkness falling, they headed back to Blue River for dinner. When they returned to the Andrews headquarters, Fred Swanson remembered that "all hell had broken loose." Seeing a huge boulder blocking one of the roads, Grant reported that the hillsides "had begun to move."[37]

Several months later, Grant told science writer Sally Duncan that there was "a dramatic power to such a huge landscape event. . . . It was absolutely the high point of my career to date, the field experience you dream about." Fred Swanson was equally enthusiastic, referring to the earth moving in the Andrews Forest, the floodwaters rearranging stream alignments and, similar to the 1964 event, bringing extensive changes to downstream rivers and riparian vegetation. On the Lookout drainage, the flood had multiple influences— debris flows, landslides, and changes to streamside vegetation. In a story for the *Oregonian*, Grant and Swanson described the multiple effects of flooding on the Andrews Forest: "landslides bulldozing forests, debris flows transporting Volkswagen-sized rocks, 100-foot logs floating down swollen creeks—in both managed and natural parts of the landscape." A preliminary assessment involving field surveys and aerial reconnaissance suggested that landslides and damage to streams were less apparent than in 1964, despite floodwaters being higher in some places.[38]

Like the 1964 flood, the 1996 downpour provided scientists with an opportunity to compare the recent event with legacies from the earlier episode. The reoccurrence of a major flood in 1996 offered scientists an opening to study how streams and fish were affected and the relation of logging practices and road construction to erosion and the severity of flooding. How much did clear-cut logging contribute to erosion and levels of flooding? Were changes in forest practices contributing to the severity of the flood? And should forest practices be made more rigorous in the light of new findings? Fred Swanson, speaking to a gathering of interested parties at the Andrews headquarters in October of that year, argued that forest management was at a critical juncture, moving in a different direction in the wake of President Clinton's forest plan. As a consequence, Swanson thought the present moment was ripe for learning and "talking to land managers about what we've learned."[39]

Because the Andrews had some of the most studied streams in the Northwest at the time of the 1996 flood, Swanson thought scientists had a wealth of information for measuring changes in streams. On Lookout Creek alone, the rushing waters moved four-foot boulders long distances, and logs roaring downstream when a debris dam burst floated off into the forest. In some areas, sediment "sandblasted" trees, and up to two feet of sediment was deposited along riparian zones. Fishery biologist Stan Gregory told the same crowd that floodwaters did not alter the stream as much as he had anticipated. Many of the expensive fish habitat restoration structures that had been constructed in the last several years had not been damaged, and fish populations seemed fairly robust. Gregory pointed out that flooding can contribute to improving fish habitat, cleaning sediment from spawning grounds and leaving small gravel that the fish require: "A flood is the best thing that can happen to fish. Biologically, fish need floods."[40]

In a lengthy article appearing in *Science Findings*, a publication of the Pacific Northwest Research Station, Sally Duncan drew on the experiences of Gordon Grant and Fred Swanson to pen a delightful article titled "Lessons from a Flooded Landscape," a summary of research from the 1996 flood. The effects of the event, according to Swanson, crystallized many assumptions already in the public domain. The rushing waters of that year created the greatest changes in small streams through debris flows. Moreover, the 1964 and 1996 events were separated by twenty-five years with very little logging in the experimental forest. In the latter flood, only 2 percent of clear-cuts and about twelve miles of logging roads were less than fifteen years old. In other words, in most of the clear-cut areas, second-growth forests were well established and, therefore, less prone to landslides. Moreover, road-building methods had been considerably improved since 1964, a factor that reduced landslides by 50 percent.[41]

Floods leave behind instructional material, Swanson argued. Although the 1964 event posed questions, most of them related to the small watersheds, the 1996 flood raised problems about entire river basins, "how material is routed through the whole system." In the second deluge, with more scientific talent on hand and a growing trove of data available, investigations focused on larger landscapes. Integrated research, Grant added, meant understanding floods as systems, because they move through entire drainage basins. The changing focus about the 1996 flood, Swanson indicates, was related to scientific inquiry shifting from ecosystems and watershed behavior to public safety after five people lost their lives in southwestern Oregon. Floods were becoming an urban-interface safety issue, "linking people with wildland hazards like fire, wild animals, and landslides."[42] Although there were other significant high-water events, the 1964 and 1996 floods remain significant markers, with legacies that still foster investigations.

Andrews scientists and other researchers continued to push for a broader understanding of ecosystem science. Fred Swanson, the lead author of a Pacific Northwest Research Station publication, reflected that the term natural variability could be associated with ecosystems existing before the coming of Euro-Americans, conditions similar to "'historical,' 'pristine,' 'prehistoric,' . . . and 'primeval.'" Managing ecosystems within their "range of natural variability," he believed, would be a scientifically sound approach to sustaining habitats for viable populations of native plants, animals, and other species. Although restoring landscapes to prehistoric conditions was impossible, there was potential in reaching an ecosystem's range of natural variability. Taking his cue from Jerry Franklin's new forestry initiatives, Swanson contended that ecosystem management centered in natural variability might achieve a "socially acceptable balance between ecological and commodity objectives." This was no quick fix, however, because it required testing "over many decades of research."[43]

Two Canadian scientists added to the push for ecosystem analysis as a new management principle. Traditional forestry and wildlife strategies had focused on limited time and spatial scales, activities emphasizing production at the expense of biodiversity. Beginning with the spotted owl, it became apparent that the single-species approach to conservation had serious shortcomings. Writing about owl-recovery efforts in British Columbia, Carlos Galindo-Leal and Fred Bunnell observed that social pressures were moving the goalposts toward larger spatial considerations, because the "species-by-species approach was clearly inefficient." Describing an ecosystem as "an area where plants and animals (including humans) and microorganisms interact with each other and with their environment," the authors viewed larger landscapes "as a mosaic of contiguous ecosystems."[44]

Ecosystem management, Galindo-Leal and Bunnell argued, was directly linked to sustainability, in which humans were critical, and where their strategies emphasized "long-term productivity, resilience to stress, adaptability to change and options for the future." Biodiversity was an explicit part of the equation. With other scientists, they believed that imitating natural disturbances was the best approach to biodiversity. The viability of species over long durations was important to the success of ecosystem management. The authors cautioned, however, that ecosystem management was always evolving, and resource professionals should continue to shape its direction.[45]

Because ecosystem science was being accepted among federal agencies during the 1990s, the Ecological Society of America published a committee report emphasizing the importance of managing human activities as well as land and water. The report traced the conventional view of humans as "lords and masters" of nature and present-day wilderness purists who argued that managing reserves and wilderness areas was unnecessary. Although scientific definitions of ecosystem management were still in limbo, the Ecological Society believed it lay somewhere between humans as masters of nature and wilderness purists. When humans were treated as part of an ecosystem, the distinction between "natural vs. unnatural breaks down"—wilderness environments were not immune from global warming, and atmospheric winds carry pollutants around the world.[46]

One of the hallmarks to the development of ecosystems as a scientifically significant field of research was the Yale University Press publication of *Ecosystem Management* in 1997. A product of a symposium at the University of Wisconsin–Stevens Point, the book included interrogations of the many-sided aspects of ecosystem management. Jack Ward Thomas's "Foreword" praised the ecosystem concept for advancing scientific knowledge about the nation's natural resources. He applauded the new approach for its expansive notion of time and scale, "because ecological processes simply do not respect jurisdictional boundaries." Thomas viewed ecosystem science as "an idea whose time has come." People might argue about its pros and cons, but there was no turning back.[47]

Addressing the management on America's national forests, Jerry Franklin's "Ecosystem Management: An Overview," was arguably the most significant contribution to the volume. Since the 1970s, scientific investigations had produced a "large body of new and significant knowledge" that had reshaped management on federal forests. Ecological research had advanced from limited temporal and spatial scales to ever-larger units of the globe over longer spans of time. Franklin praised the applicability of ecosystem concepts to larger spatial scales as one of its strengths. The International Biological Program's biome studies contributed mightily to advancing research and knowledge about ecosystems, especially the pioneering investigations at Hubbard Brook Experimental Forest.[48]

Franklin applauded the Greater Yellowstone Ecosystem concept for alerting ecologists to the limitations of restrictive boundaries for places where wide-ranging wildlife species roamed at will. That discovery led to new views of landscape ecology and the ecological health associated with larger, contiguous spatial patterns. The truth to those findings was reflected in the expansive migratory habits of elk herds, neotropical birds, sea mammals, and anadromous fishes. Franklin also offered insights to sustainability—maintaining "the potential of our terrestrial and aquatic ecosystems to produce the same quantity and quality of goods in perpetuity." Those objectives could be accomplished through flexible, "evolutionary" management policies that recognized the dynamics of ecosystems.[49]

Yellowstone National Park served as a national laboratory for ecosystem studies, a place that naturalist Paul Schullery described as "a kind of perceptual experiment that will never end in our quest to understand this exasperatingly elusive thing we call nature." Historian James Skillen argued that Yellowstone was the home ground for federal inquiries into ecosystem management, the place where the benefits of using ecological methods were first realized, and where some of its most implacable problems were confronted. The Yellowstone experience also revealed the intense politics involving anything that challenged the status quo. Skillen added that a broader understanding of ecosystem management can be gleaned from investigations into elk, bear, and wolves, emphasizing the need to look beyond the park's boundaries for appropriate habitats.[50]

Gene Likens, a respected ecosystem scientist whose research centered on long-term studies at Hubbard Brook Experimental Forest, described ecosystem ecology as an effort to develop a systematic understanding of the natural world and to seek solutions to environmental problems. In a presentation to the Institute for Ecosystem Studies in 1998, he pointed out that ecosystem science provided a way to capture "the linkages and complexity" of the components of nature. Despite the gloomy writings of Alston Chase, he thought the ecosystem field was enjoying a "somewhat giddy stage" in its intellectual development, where proponents were asking whether they were "'smart enough' to pose and then tackle the extremely complex and multifaceted questions that are generated at the ecosystem level of organization." Because the study of ecosystems offered no simple answers, Likens wondered if scientists were asking the right questions. Were there optimal sizes to ecosystem studies, and how should one determine the proper scale? And what about interpreting data at different temporal scales? Data from Hubbard Brook for periods of one to five years, he reported, was misleading when one considered longer periods of time.[51]

It was critically important for ecological research to be integrative, Likens insisted, to have its work "lie at the intersection of disciplines." There were few generalists in ecosystem studies, because it was difficult to master both aquatic and terrestrial ecosystems. "Multidisciplinary, team approaches," therefore, "is common now (if not required) in such large National Science Foundation–funded projects as Long-Term Ecological Research." (The Andrews Experimental Forest had been carrying on team investigations since 1970.) Experimental manipulation, Likens thought, was important to ecosystems ecology, where scientists could study entire environments such as lakes and watersheds. Experiments in watershed ecosystems existed in only a few places: Hubbard Brook in New Hampshire, Coweeta in North Carolina, and the Andrews Forest in Oregon. Likens cited the void between ecosystem science and the implementation of such policies, especially in the United States.[52] With Republicans gaining control of the House of Representatives in the 1994 election and the ascension of Newt Gingrich to Speaker of the House, ecosystem studies would be subject to increasing criticism.

Politics, always important to scientific inquiry, became increasingly testy when investigations involved land and water resources. When Bill Clinton was elected president in 1992, both the executive and legislative branches of government were controlled by the Democratic Party. In addition, Clinton appointed people with established environmental credentials to key natural resource positions: Bruce Babbitt as interior secretary, Jim Baca to direct the Bureau of Land Management, Mollie Beattie to head the US Fish and Wildlife Service, Carol Browner as EPA administrator, Jack Ward Thomas as chief of the Forest Service, and many others as assistant officials. The question going forward was whether advances in ecosystem science during Clinton's two terms in office would endure. Congressional Republicans, generally suspicious of policies related to ecosystem management, made life difficult for ecologists when they gained control of the House in 1994 and the Senate in 1998, leaving the Democratic president to fend off Republican initiatives.[53]

Despite the Gingrich congresses and Republicans winning the contested presidential election in 2000, many federal and state natural resource agencies had adopted ecosystem approaches. Working mostly beyond the public's notice, land managers had been introducing ecosystem policies, threatened only when resource industrialists accused the program of destroying jobs and hurting rural communities. Events in the early 1990s—Forest Service chief Dale Robertson declaring ecosystem management the agency's official policy, President Bill Clinton's forest summit in Portland, and the exhaustive FEMAT report—helped bring ecosystem management to public attention. Citizen reactions to media coverage wavered depending on local circumstances—the health of local economies and the effectiveness of environmental appeals. David Bengston, with the Forest Service's Northern Research Station in St. Paul, Minnesota, found that media accounts during this period were generally favorable toward support for environmental and ecological values.[54]

Among those thoughtfully reflective about the ecosystem concept during this period was Robert O'Neill, a scientist with the Oak Ridge National Laboratory in Tennessee. O'Neill directed his strongest criticisms at those who saw ecosystems as machine-like natural units in which every item was subject to precise measurement. He was skeptical that "the enormous complexity of natural systems" was suited to computerized models, because the ecosystem concept was "an a priori intellectual structure, a specific way of looking at nature." It addressed some of nature's properties and ignored others. Its limitations, which were apparent to many, invited a backlash among those who saw ecology "as unnecessarily constraining human freedom and economic growth." O'Neill attributed some of the opposition to the "apocalyptic fervor" of environmentalists, who bought in to the idea that ecologists were predicting doom and gloom.[55]

Ideological critics, on the other hand, charged that ecosystem science was hindering the development of the human prospect, restricting what people could do in the biophysical world. There was some substance to those criticisms, O'Neill contended, because classical ecologists had been emphasizing the stable, self-regulating balance of nature. Despite such misplaced arguments, O'Neill positioned the ecosystem idea as a paradigm, "a convenient approach to organizing thought," a human construct providing a way to understand the complexity of the physical environment. Even then, he noted that ecologists liberally used ambiguous terms with great conviction.[56]

Those admonitions aside, O'Neill thought there were assumptions in those discussions worthy of exploration. The ecosystem concept was useful as "an explanatory framework for ecological phenomenon." With all its weaknesses in definition and uncertainty, it provided some understanding of ecosystem theory. He described an ecological system as "a range of spatial scales, from the local system to the potential dispersal range of all the species within the local system." In the end, the degree of human disruption would determine an ecosystems' sustainability or collapse, because *Homo sapiens* were "a keystone species within the system." In an earlier publication with a colleague, O'Neill cautioned that much of the current work in ecology failed to consider humans as a keystone species, positing humans as external disturbances in nature. That approach erred because humans were, "in fact, another biotic species within the ecosystem and not an external influence."[57]

Republican control of Congress in the late 1990s hampered progress on the expansion of ecosystem science in federal agencies. Control of the legislative branch, however, had little success in restricting the influence of environmental laws. One legislative measure, a rider to the 1995 Rescissions Act, allowed sales of fire- and insect-damaged timber on federal land and prohibited citizen appeals. The work of Senators Mark Hatfield of Oregon and Brock Adams of Washington, the measure outraged environmental organizations, because some of the logging took place in old-growth forests. When the dust settled from the fractious rider, its supporters had little appetite for renewing it, and loggers had harvested less than 1 percent of the remaining old-growth stands in the Pacific Northwest. Environmental organizations were also successful in stopping Forest Service timber sales when the agency attempted to bypass the Northwest Forest Plan without conducting required surveys.[58]

In the midst of the stalemate between executive and legislative branches, Forest Service chief Mike Dombeck, who succeeded Jack Ward Thomas in 1997, sought to implement ecosystem practices in the agency's mandate to protect watersheds in the national forest system. To promote those objectives, Secretary of Agriculture Dan Glickman appointed an interdisciplinary Committee of Scientists to review the directive: "Biological diversity, use of ecosystem assessments in land and resource management, planning, spatial and temporal scales for planning, public participation processes, sustained forestry, interdisciplinary analysis, and any other issue that the Committee identifies that should be addressed."

Secretary Glickman's Committee of Scientists began a cross-country tour in early 1998, visiting Forest Service officials, tribal representatives, other federal resource agencies, and state and county governments. K. Norman Johnson, of OSU's College of Forestry and Gang of Four renown, chaired the thirteen-member committee, whose directive was to recommend how to promote resource planning that met the requirements of the nation's environmental laws. Committee personnel with some experience on the Andrews Experimental Forest included Robert Beschta, an OSU forest hydrologist, and James Agee, a forest ecologist from the University of Washington.[59]

The *Journal of Forestry* published Johnson's summary report in May 1999, emphasizing sustainability as the mission of the national forest system and ecological sustainability as the foundation for stewardship. To accomplish those goals, the committee proposed the need to maintain ecological integrity, to preserve habitats for threatened and endangered species, and to carefully monitor the work. Johnson acknowledged the variability of ecological systems "and our incomplete knowledge of them," recommending that planning "should operate within a baseline level of protection for ecological systems and native species." With a presidential election looming on the horizon one year hence, Johnson recognized that congressional action or inaction and administrations "can undercut plans and render collaborative planning frustrating and ineffective."[60]

George Hoberg, of the University of British Columbia, indicated that the Committee of Scientists "never made a scientific case for the primacy of ecological sustainability." Before the Clinton administration left office, however, it proposed a new national forest planning rule committing the Forest Service to ecologically sustainable forestry. The rule embodied "the principles of multiple-use and sustained yield without impairment of the productivity of the land." Clinton's successor, President George W. Bush, immediately moved to rescind ecological sustainability in lieu of economic and social priorities, including the requirement for protecting species. The new president's revised rule included permissive phrasing such as "to the extent possible." Like many issues with ecosystem management, the lack of clear statutory authority allowed the new administration to make changes at will.[61]

By the twenty-first century it was clearly apparent that ecosystem management had become a political football, waxing and waning with the priorities of successive presidential administrations. Through the 1980s and 1990s, attention to healthy ecosystems gained popularity as scientific evidence pointed to environmental problems and losses of biodiversity, and by the mid-1990s, natural resource agencies began adopting ecosystem programs. Although the Forest

Service and BLM had committed to ecosystem strategies, congressional pressures, focusing on the production of material goods, hampered their ability to
incorporate ecosystem practices into management priorities. To fully embrace
ecosystem practices required new ways of thinking about complex concepts
that meant different things to different people.

Implementing ecosystem policies in the field has proved a minefield for
federal land-management agencies. Despite scientific evidence to the con-
trary, administrations had a propensity to manage for single species, ignoring
the health of larger ecosystems. Federal managers continually faced congressional demands for a steady and reliable flow of goods, giving officials little
incentive to place their careers in jeopardy to pursue holistic management
strategies. Although scientists largely agreed that ecosystem management
required scientific knowledge, legal and structural restrictions limited their
work. The bifurcated ownership of landscapes into discrete units restricted the
legal authority to manage properties across ownership boundaries. Changes
in political power and funding proprieties also opened or closed opportunities for ecosystem approaches. Administrators and field personnel in the BLM
and Forest Service agreed that political barriers presented the most important
factors in managing their lands.[62]

The incoming George W. Bush administration immediately set about
reversing the Forest Service Roadless Area Conservation Rule put in place
before Bill Clinton left office in January 2001. A spin-off from the Northwest
Forest Plan, the ruling prohibited building roads through inventoried roadless areas totaling more than fifty-eight million acres in the national forest system. The new Republican administration viewed the ruling a violation of the
National Forest Management Act, because it closed areas to timber harvesting. Unwilling to directly challenge environmental protections, Bush's natural
resource policymakers pushed modest environmental initiatives, including
the Healthy Forest Restoration Act passed in 2003. James Skillen observes
that Bush put a different spin on forest health: "Rather than measuring forest
health in terms of biodiversity or old-growth characteristics, the Bush administration measured it in terms of timber production and wildlands fire risk."[63]

In rejecting Clinton's ecological principles, Bush policymakers continued
to use the rhetoric of collaboration, promoting "cooperative conservation,"
which under Interior Secretary Gale Norton featured a libertarian model of
conservation, leaving management to local participants who "live and work
on the land." Under those directives, ecosystem management morphed into
a conservation mode that Skillen claims was "stripped of many substantive

ecological commitments."[64] Lacking definitive clarity, the meanings of ecosystems and ecosystem management proved adaptable to different interpretations under succeeding presidential administrations.

Aside from the lack of support for ecosystem management at the federal level, Skillen observes that ecological science has continued to influence management across broad sections of land in the American West and beyond— "agencies simply cannot put resource management questions back into a simple formula of sustained yield or rectilinear boundaries of national parks or national forests." Because of existing environmental laws, natural resource agencies must bend to the will of public accountability and prevailing scientific norms. The implications of ecosystem initiatives during the Clinton administration, therefore, marked a redirection, if modest, in federal land-management policies that include input from a wide range of jurisdictions and interests.[65]

Despite naysayers and the inherent complexity of ecosystem science, the Andrews Experimental Forest was fully committed to biodiversity and long-term studies that fully embraced interdisciplinary approaches to all components of the physical environment. With the continuing support of the National Science Foundation, researchers extended their studies into analyses of fires, windstorms, and debris flows, linking the Lookout Creek drainage to other river basins in the western Cascades. Forest and stream stewardship, critically important to land managers, were important topics of communication between scientists and land-management agencies. LTER grants in the twenty-first century increasingly featured invasive species and climate change, issues of common concern among Forest Service and BLM officials. Forestry personnel were interested in the effects of warming temperatures on forest/ stream interactions and the socioeconomic consequences for citizens living downstream. The social sciences, in effect, joined with the ecological sciences in expanding the boundaries of inquiries on the experimental forest. In the second decade of the new century, declining snowpacks, earlier spring seasons, and autumn extending through October affected directly and indirectly residents in the Willamette Valley. Through all those changes, Andrews personnel continued to monitor updates to the Northwest Forest Plan and the Forest Service's efforts to implement its provisions.

Chapter Six
Biodiversity and Long-Term Studies

The scientific community is united in the view, informed by a body of evidence amassed over more than 50 years, that climate change caused by humans poses a considerable threat to life here on Earth.
—Richard Hudson[1]

As the twenty-first century proceeded apace, H. J. Andrews scientists were moving forward on a broad front of activities—biodiversity, carbon and nitrogen dynamics, climate issues (variable and rising temperatures), human and natural disturbances (forest practices, fire, wind, landslides, and floods), and hydrological research (vegetation influences on stream temperatures), forest/stream interactions, and vegetation (regeneration and mortality using long-term plot studies). Their ambition was to provide land managers and policymakers with guidance for making decisions. In the Andrews annual report to the National Science Foundation (NSF) in 2000, they linked their work to the critical question—"How do land use, natural disturbances, and climate change affect three ecosystem properties: carbon dynamics, biodiversity, and hydrology?"[2]

By the turn of the century, the Andrews headquarters had transitioned from secondhand trailers to sophisticated field research facilities equipped to house visiting scientists in small apartments, a dormitory, and a meeting hall to host daylong symposia. The Quartz Creek and Roswell Ridge apartment buildings (built in 1992 and 1993) have multiple bedrooms and baths capable of housing thirty-four occupants; the Rainbow apartments (constructed in 1994) have the capacity for eighteen people. A custodian's residence, completed in 1989, comprises a remodeled trailer with nine hundred square feet of space; a seven-hundred-square-foot director's cabin has a shed roof, bedroom and bath, and a living/dining area. Other offices and facilities include a 5,400-square-foot laboratory/office building (1993); a 4,800-square-foot education building (1999), with an auditorium capacity of one hundred seats; a laboratory classroom (capacity thirty); and small offices and a kitchen. An

Entrance to H. J. Andrews Experimental Forest. Photo by Tom Iraci, April 2005.

open-sided pole shelter, the Salt Salmon Pavilion (1993), with a slab floor, electrical power, and available water, is capable of seating sizable numbers of visitors. A variety of funding sources over the years have made possible the expanding facilities—congressional earmarks, Forest Service special and competitive grants, and National Science Foundation grants.[3]

Accessory facilities are arrayed around the perimeter of the headquarters: a gray, uninsulated thousand-square-foot barn (1973) with a small insulated office; a pump house (1982); a four-bay, drive-through, two-thousand-square-foot snowcat shop (1988); a small gas house (1994); and an open-sided recycling shed (1995). Beyond the headquarters, a small shelter at the main entrance (1989) protects a kiosk providing cover for posting informational material and guiding visitors to the compound. The compound known as the Gypsy Camp includes eight uninsulated cabins, each one hundred square feet, set on blocks without electrical power or water. Four small insulated cabins on Mack Creek (1980), McRae Creek (1983), Carpenter Mountain (1987), and Wildcat Mountain (1987) provide year-round shelter for researchers working far from the headquarters location. (Wildcat Mountain is a thousand-acre Research Natural Area located in the Willamette National Forest north of the Andrews Forest.)[4]

The modern facilities vastly expanded the ability of the Andrews to host visitors from near and far and to offer NSF-funded training programs through the

H. J. Andrews headquarters, snowy scene. Photo by Al Levno, January 1993.

Northwest Center for Sustainable Resources at Chemeketa Community College for schoolteachers and community college students. Site director Art McKee was a central player in many of the outreach programs until he retired in 2002. Oregon State University and NSF funded a summer Research Experience for Undergraduates (REU) that brought talented students from across the nation to engage in hands-on investigations of stream and terrestrial investigations in the Lookout Creek drainage. The experimental forest also participated in Oregon State University's annual SMILE program (Science and Math Investigative Learning Experiences), an initiative to provide Hispanic, Native American, and other minority youth the opportunity to expand their knowledge of science, mathematics, and health-related issues. And, beginning in 1989, the Andrews began hosting an annual field day for people internal to the program, the interested public, and state and federal natural resource administrators.[5]

After several years serving as the Andrews lead principal investigator (PI) with NSF, Fred Swanson ceded the position to Mark Harmon in 1989, the first OSU faculty to hold the position. In its annual Long-Term Ecological Research (LTER) report to NSF for 2000, project administrators announced a new governance plan, including an executive committee, with one junior scientist serving for one year to develop leadership experiences for the future. One purpose of the executive committee was to seek greater support from the university for its LTER program, a major recommendation of NSF's midterm review committee. The governance strategy took note of the generous support of the Pacific Northwest Research Station, with lesser contributions from OSU. The new management policy involved a series of "working committees" to provide leadership experiences for potential principal investigators for the administration of the LTER.[6] (For a listing and the dates for the LTER grants and the principal investigators, see appendix A.)

Disturbance analysis, a mainstay of research on the Lookout Creek watershed, was an important component in most LTER reports. In its 2001 submission to NSF, Andrews personnel reported investigating the occurrences of forest fires, and damages from high winds, landslides, and flooding—studying "retrospective histories of fire, debris flows, interpretations of flood effects, and general inter-site comparisons on disturbance." One principal event, the February flood of 1996, offered scientists the opportunity to explain how large floods affected forested landscapes. Subsequent investigations revealed the effects of high volumes of water flowing through a watershed, "the role of moving wood as a disturbance 'tool,'" and the shape and "importance of refuges in the stream/riparian network." A graduate student contributed to the disturbance studies with investigations on fire histories in western Oregon, climatic conditions, and the effects of fire on vegetation. A researcher from the University of Vermont added insights to relations between riparian and upland fire disturbances on forest stands.[7]

Through this period, Andrews personnel carried on important liaisons with regional land-management agencies, the National Park Service, the US Geological Survey's Biological Resources Division and its Water Resources Branch, and the Forest Service. Collaborative work involved comparative plot studies and collecting data on small streams. They engaged in inter-site activities, with fourteen senior scholars and twelve graduate students taking part in an LTER All Scientists Meeting in Snowbird, Utah, in August 2000. Scientists also participated in collaborative enterprises with international LTER sites in Australia, Canada, China, Japan, New Zealand, South Africa, Russia, and several European countries. Fred Swanson continued his long association with Japanese scientists, and Mark Harmon was involved with a collaborative project in northwestern Russia.[8] At the beginning of the new century, the Andrews Experimental Forest was a busy place, especially during the summer months, carrying on professional outreach with the public and public agencies in the region and important links with global scientists.

Since the 1950s, the experimental forest had been deeply involved in scientific investigations of forests, streams, and other issues important to land managers.

Scientists from multiple disciplines had been studying natural and managed forests, inquiring into climatic conditions, forest/stream relations, succession patterns in vegetation, natural and human disturbances, biological diversity, and carbon and nutrient dynamics. Scientists had ranged far afield to places like Mount St. Helens to examine recovery from the volcanic eruption of 1980. Interdisciplinary research in long-term ecological programs involved investigations across vast temporal and spatial scales. From these ecosystems inquiries, Andrews scientists had made significant contributions to forest and stream stewardship, arguably positioning themselves as models for scientific contributions to policymaking.[9]

Since the launch of NSF's Long-Term Ecological Research program in the United States, sites have been part of the NSF-created LTER Network. Although NSF established the network, it was not a top-down enterprise, because the agency provided no funding. By the 1990s, the network was serving as a gathering place for cross-site network-wide data, and information about international studies. The largely volunteer network searched for general ecological principles that could be identified in many ecosystems at different levels. The US LTER Network numbered twenty-one sites in 1998, representing quality research, strong data sets, and a commitment to long-term investigations. The network gathered information on general ecological trends over long periods of time, covering vast land- and waterscapes and to create "a legacy of well-designed and documented long-term experiments." Equally important, it was to respond to environmental issues important to society. In addition to its central office, the network had a coordinating committee made up of representatives from all the member sites.[10]

During the twenty-first century, the Andrews Forest has continued to operate in partnership with Oregon State University, the Pacific Northwest Research Station, and the Willamette National Forest. Its participation in science-related policies on the region's national forests linked university scientists with Forest Service researchers in multidisciplinary watershed-wide investigations important to advancing scientific knowledge and management practices. Long-time scientists with the Andrews provide a living memory dating to the 1960s, with recollections involving political cycles, the ups and downs of the national economy, and the introduction of new technologies and research venues. Through it all, they have coordinated their research findings with participating LTER sites on topics related to "water, climate, vegetation, animals, and use of natural resources."[11]

In its application for a continuing six-year NSF grant in 2002, the Andrews proposal underscored LTER funding as "the primary meeting ground for managing much of the entire enterprise," including activities on Research Natural Areas and other experimental forests in the Northwest. Preparing submissions to NSF was exhausting, including covering the full breadth of activities on the experimental forest and adjacent sites, a summary of previous work, publications, a description of investigations for the next grant period, program management, and an outreach agenda. The "Project Summary" for the 2002 grant committed Andrews researchers to important guiding questions: "How do land use, natural disturbances, and climate change affect three sets of ecosystem services: carbon and nutrient forces, biodiversity, and hydrology?" Those factors, the proposal stated, were principal drivers of change in the region. The approach would be retrospective, using temporal observations across the past five hundred years, experiments, and a promise to continue long-term studies.[12]

The 2002 submission to NSF (LTER5) addressed research partnerships under the Northwest Forest Plan—the Central Cascades Adaptive Management Area and the Blue River Landscape Study—using "historic disturbance regimes to set frequency and severity of forest harvest/cutting." The Blue River plan was "a form of science synthesis, whereby scientists from diverse fields help in formulation, plan development, and monitoring." Although the LTER provided little support for the Blue River project, Andrews scientists supplied the basic information about ecosystems research related to streams, riparian areas, watersheds, and various forested landscapes.[13]

A new principal investigator, Oregon State University scientist Barbara Bond, authored the concluding report for LTER5. In a riveting summary, she noted that Andrews research had now passed through multiple funding cycles involving "climate variability and long-term change, land use, and natural disturbances," the forces altering the environment. The Andrews Forest and the Blue River watershed's 55,000 acres were the setting for much of the research. Small watersheds, a special focus of LTER5, were critical landscapes for studying climate, ecosystems, and hydrological processes such as stream flow, nutrient transport, and changes in vegetation. For the duration of the grant period, Bond highlighted new leaders, including her election as principal investigator following Mark Harmon, who had stepped down. She explained the forest's governance strategy of having orderly leadership transitions midterm for each funding cycle. Forest Service stream ecologist Sherri Johnson succeeded Fred Swanson as lead scientist representing the Pacific Northwest Research Station

Tom Spies has spent a long and productive career as a forest ecologist with the Pacific Northwest Research Station in Corvallis. Spies served as the overall leader for a major Forest Service study of the Northwest Forest Plan in 2018, *Synthesis of Science to Inform Land Management within the Northwest Forest Plan Area.*

in LTER leadership, and Mark Schulze was hired as the new forest director of the Andrews site and Lina DiGregorio as research coordinator.[14]

Among the cardinal accomplishments of LTER5 was its emphasis on biodiversity research devoted to how native plant communities influenced exotic invasions and the special traits of the invading plants that enabled their successes. The presence of Lepidoptera (butterflies and moths) from eastern Oregon reflected the changing climate. Addressing invasive species would be an important research agenda in LTER6. Long-standing research on disturbances continued, with ongoing investigations of fires, floods, spruce budworm defoliation, and "snowdown" (heavy snow toppling trees). Studies of historic wildfires reflected complex patterns of regeneration depending on topography, microclimates, and biotic factors. Investigations of the 1996 flood disturbances to streams indicated few changes to small channels, but in larger channels downstream, large pieces of wood rearranged waterways.[15]

The LTER5 disturbance component included an update on the Blue River Landscape Study, reporting an "on-the-ground application of the concept of using history to design future forest landscape management with a view stretching over 500 years into the past and several centuries into the future." Although "regional forestry politics" had stalled the Central Cascades Adaptive Management Area timber sales on the Blue River study, scientists continued to see value in the project. Andrews personnel proposed a solution

Barbara Bond, an Oregon State University professor of forest science, was the only woman to serve as principal investigator for the experimental forest's Long-Term Ecological Research program (2008–2014). Photo by Cheryl Hatch.

involving social scientists who canvassed local communities about using history to design future management practices. The surveys revealed "that the public has a hard time comprehending the ideas because our vocabulary is not clear." The LTER report, however, defended the strategy of using the dynamics of historic landscapes to guide forest management.[16]

The submission of the Andrews LTER6 application in February 2008 was its most elaborate and detailed to date. Under principal investigator Barbara Bond, the proposal outlined an ambitious program of long-term experimental activities, some of them representing accumulations of more than fifty years of data. The submission committed the experimental forest to long-standing forest/stream investigations and evaluating ecosystem responses to climate change. Rising temperatures in the next half century, the document cautioned, would far exceed increases of the previous fifty years. Given those expectations, the experimental forest was well positioned to assess anticipated environmental changes, because it possessed "measured and proxy records extending to 50 and > 500 years respectively, and our site spans steep and complex climate gradients."[17]

Reflecting previous proposals, LTER6 would draw on studies of biota and ecological and geophysical developments at research sites on the western and eastern slope of the Cascade Range—other experimental forests, Research Natural Areas, and long-term plot studies. The Andrews would continue its

leadership role encouraging inter-site scientific exchanges among LTER
Network members. Another major LTER6 initiative involved integrating
the social sciences and the ecological sciences to bring the program to local
citizens. Collaboration with other organizations was necessary, because NSF
funding would not cover such initiatives. Social networking would include
field demonstrations of management strategies projecting the influence of climate change on landscapes. "Alternative futures," the proposal declared, "will
help communities and institutions make choices."[18]

An important component of public outreach under LTER6 was an
increased emphasis on how "ecological services" from natural resources on
public lands benefited citizens. Ecological services included "carbon sequestration, wood production, water supply, and habitats to support biodiversity."
While some services were local, the extent of others, such as carbon sequestration, was global. It was important, however, for social scientists to assist
in taking the issue of ecological services to public audiences. A related effort
was the Long-Term Ecological Reflections program, involving environmental
writers, philosophers, and artists, who would explore the multifaceted natural
world and research on the Andrews. Their objective was "to examine how the
senses of awe, hope, environmental ethics, and other points raised by humanists play out in different environmental and social contexts." Creative work in
the humanities would reach citizens who otherwise had no knowledge of the
Andrews Experimental Forest.[19]

Another initiative originating in 1997 involved annual symposia (although they
haven't been held strictly every year) to share personal research activities and to
foster camaraderie among the growing number of scientists affiliated with the
experimental forest. The First Annual LTER Symposium featured eleven presentations, Mark Harmon's opening address, Fred Swanson's "Overview," and
a poster session with twenty-one posters. The symposium in 2000 dropped
the LTER reference and used a thematic phrase, "Future Ecosystem Science
at the Andrews," to characterize the presentations. Subsequent symposia followed a similar script, with more elaborate descriptions: New Adventures
in LTER Science: Transitioning from Spatial to Temporal Studies (2002);
Carbon, Nitrogen, and Water Interactions in a Forested Ecosystem (2003);
New Ideas and New Directions for Long-Term Ecological Research: Building
on Our Legacies (2007); and Networks and Synthesis (2010).[20]

The year 1998 marked the fiftieth anniversary of the H. J. Andrews
Experimental Forest. To commemorate the five decades since the forest's

founding as the Blue River Experimental Forest, leaders of the cooperating partners—Fred Swanson (Pacific Northwest Research Station), Art McKee (Oregon State University), and Lynn Burditt (head of the Blue River Ranger District)—began planning a series of events that would extend through the year. Forest director McKee arranged public tours of the forest for May 16 and another on September 12 involving field stops and discussions about old-growth forests, stream ecology, the log-decomposition study, and landslides ("Geomorphology of Earth Flows"). On the second tour at the headquarters site, Art McKee's talk was titled, "Lessons from the Ancients: What We Learn from Old-Growth Forests." The Eugene Natural History Society sponsored McKee's address, as well as three others on the University of Oregon campus, all of them to honor the fiftieth anniversary of the Andrews Forest.[21]

The principal celebratory event of the fiftieth anniversary was a one-day gathering at the Andrews headquarters on August 21. The theme, "Celebrating a Legacy of Cooperative Science and Management," featured Art McKee and Lynn Burditt welcoming nearly three hundred people who came for a day of tours around the headquarters site and to experience tree climbing in old-growth Douglas-fir next to the pavilion. "Speeches, workshops, and reminiscing" were highlights of the event. Limited to ten-minute presentations in the morning, speakers included OSU president Paul Risser, Art McKee, Lynn Burditt, and Fred Swanson; Richard Guldin, director of science policy, US Forest Service; Bruce Hayden, division director, National Science Foundation; Jerry Franklin, University of Washington; and Charles Burns, grandson of H. J. Andrews.[22]

Following lunch, concurrent thirty-minute sessions filled the rest of the afternoon, with speakers addressing past, present, and future research in locations around the headquarters site. The presenters, all well-known scientists, addressed familiar research topics associated with the Andrews: old-growth forests, debris flows in mountains, ecology of belowground organisms, young-stand management, floods and stream ecology, carbon dynamics and climate change, people in the environment, ecology of the northern spotted owl, landscape ecology projects, history, the Andrews, science and community, ecology of canopies, and biological diversity, in addition to other activities such as landscape and nature photography and exhibits of artwork, historic photographs, and notable publications."[23]

For the tenth symposium in 2009, Andrews organizers focused on the sixth LTER grant, "Using Innovative Approaches and Long-Term Research to Address Complex Socio-Ecological Questions." A single-day event, the gathering featured OSU scientists Kate Lajtha, Barbara Bond, and Chris

Daly addressing watersheds, carbon cycling processes, and climate change. Denise Lach, an OSU sociologist, spoke to her specialty, "the natural/human interface." The Forest Service's Sherri Johnson presented a paper on biotic responses to early warming in the spring, and Tom Spies described, in "Digital Forest," how the Andrews Forest appears through an alternative lens. OSU faculty Julia Jones and Mark Harmon filled out the morning agenda with talks on using retrospective analysis and modeling as ways to learn about potential futures. A poster session following lunch illustrated recent research findings and offered a fascinating array of investigations—nitrogen changes in old-growth forests, ecosystem production, how models assist in understanding ecosystems, groundwater in watersheds, wireless communication on the Andrews, and other intriguing projects.[24]

The 2010 symposium, "Networks and Synthesis," offered a series of morning presentations, with the afternoon devoted to a poster session. Because the Andrews Forest had been preeminent in networking across LTER sites—one of the founding objectives of the National Science Foundation—Mark Harmon led discussions of network experiment successes and how those achievements carried forward into the future. Denise Lach followed with intra-site studies of vegetative cover, changing patterns of land use, and how local ecological knowledge worked among LTER sites. Fred Swanson praised the emerging role of the humanities and arts in enriching knowledge about the Andrews and other sites. For the synthesis component, Sherri Johnson reviewed cross-site data banks and their value in understanding comparative historical change among different LTER sites. Lydia Ries O'Halloran spoke to her investigations of nutrient limits in grasslands across the LTER network, and Tom Spies summarized the "challenges and opportunities" in collaborating among sites in future research projects.[25]

The annual symposia continued into the next decade, with a common format of presentations on research and poster sessions. The one persisting theme in the annual events has been long-term ecological research and the interactions of complex phenomenon, terrain, aquatic environments, climate change, and the projections of those inquiries into the future. "Thirty Years and Counting," Mark Harmon titled his opening talk in 2015, describing his "200-year log decomposition experiment." He opened the symposium the following year with a talk titled "The New Abnormal," explaining what scientists learn from long-term observations of vegetation. Brian Black, a visiting scientist from the University of Texas, was up next, with a paper assessing eight centuries of environmental change in the forests of western Oregon.[26]

Sherri Johnson, a Forest Service stream ecologist with the Pacific Northwest Research Station, succeeded Fred Swanson as lead scientist representing the Pacific Northwest Research Station in LTER leadership.

The annual Andrews symposia have offered something for everyone, with the announcements always extending invitations to the public.

In a review of the Andrews Experimental Forest in January 2018, Fred Swanson offered a brief summary of its investigations from 1994 to 2010 and from 2010 to 2018. His notes reflect intense research-management partnership activities during the attempts to implement the Northwest Forest Plan. The Blue River Landscape Study was an exercise in using data gathered from historical disturbances to shape new management strategies and silvicultural experiments that would produce wood fiber and advance old-growth characteristics in forest stands. Swanson's assessment for the years since 2010 reveals a dramatic curtailment in cooperative partnerships because of declining support for adaptive management strategies. Although the breadth and depth of management research options has declined, basic science has continued, including a few adaptive management enterprises and inquiries into "evidence of a strong warming signal in the temperature records."[27] Although there were roadblocks to some of the more ambitious research opportunities under the Northwest Forest Plan, Andrews scientists could draw on some sixty years of records and accumulated data going forward.

One of the problems in pursuing adaptive management harvesting strategies, or what environmentalists sarcastically referred to as "wild science," reflected practices that challenged the social norms of conventional forest investigations. In a draft paper, Hannah Gosnell and three colleagues (including Swanson) argued that the purpose of the Augusta Creek project and Blue River Landscape Study was to experiment in emulating past disturbances, using targeted harvests to build greater diversity and resilience in forests. Environmentalists, who zealously opposed harvesting *any* trees, took conservative, even reactionary positions, dubbing such proposals as simply another Forest Service excuse to cut more trees. Andrew Gray, a proponent of experimentation, viewed opposition to experiments as a severe hindrance to advancing science.[28]

In their draft, Gosnell and colleagues addressed the tension in forest-management proposals between those who favored leaving forest lands alone and others who believed landscapes should be managed to maintain resilience in the face of the increased danger of wildfires from a warming climate. Historical research on the experimental forest supported both positions. The initial successes were investigations leading to the northern spotted owl decisions and the preservation of old-growth forests. More recently, under the Northwest Forest Plan, scientists supported active management, "harvesting in older 'matrix' stands and in promoting experimentation in older stands in Adaptive Management Areas."[29] Despite disappointments in their inability to pursue such research, Andrews scientists still enjoyed a full plate of research opportunities in the second decade of the twenty-first century.

Discussions about climate change among Andrews scientists dated to the 1980s, when researchers began addressing the "greenhouse effect," the consequences of atmospheric gases trapping heat radiating from the Earth's surface. As early as 1983, G. M. Woodwell et al. linked global deforestation with rising levels of carbon dioxide in the atmosphere. Four years later, James Hansen and Sergei Lebedeff published an article pointing to increased surface air temperatures across the globe. The widely read *Scientific American* pointed to a consensus among climatologists in 1989 that the world was warming, glaciers were melting, and sea levels were rising, trends that were expected to continue into an "indefinite future," posing dire threats to agriculture, forested regions, and water supplies. Respected climatologist Stephen Schneider added that atmospheric scientists firmly believed that increasing concentrations of carbon dioxide and other gases would accelerate "heat trapping and warm the climate."[30]

Writing for *Northwest Environmental Journal* in 1990, lead scientist David Perry of the Department of Forest Science at OSU and an Andrews LTER scientist, observed that greenhouse gases were accumulating and that warming temperatures would "affect plant physiology through increased respiration relative to photosynthesis, longer growing seasons and effects on the yearly developmental cycle." Evidence indicated that natural disturbances, insect and disease infestations, wildfires, and dramatic swings in precipitation would increase. Fred Swanson and colleagues followed with evidence that a warming climate would have a significant effect on watershed science and forest management. The Corvallis office of the Environmental Protection Agency added its voice to threats that rising temperatures posed to Northwest forests in 1992—the region's woodlands could change from one vegetation type to another, and "forest disturbances such as fire, wind, and pest/pathogen outbreaks will likely increase."[31]

At a climate workshop with the California Department of Water Resources in 1993, University of Oregon geographer and Andrews scientist David Greenland reported on historical climate information gathered at the Andrews Experimental Forest since 1951, the data gleaned from the primary meteorological station at the headquarters site and a series of temperature and precipitation recording sites scattered across its 15,800 acres. Those records, he argued, were important to positioning climate findings on the Andrews in a regional context. Working in part from average temperature and precipitation data collected between 1973 and 1991, Greenland reported increased maximum and mean temperatures, with the most significant increases occurring between March and May. Using those figures, he concluded that the Andrews climate was representative of the western Cascades and the Pacific Northwest, and that the forest was "well coupled with these hemispheric-scale events."[32]

Given the persisting skepticism of a small percentage of the public about human-caused global warming, it is striking to note the scientific community's overwhelming sense of confidence, even then, that climate change was real, that humans were the principal cause, and that it was generating noticeable changes in the biosphere. In a Spring Creek–sponsored event,[33] Michael Nelson and Kathleen Dean Moore joined other authors from the Columbia River Forum in addressing a "strong and rapid societal response, especially in the US" to the warming climate. Their plea, published in 2009, came in the midst of a frustratingly slow public reaction to what they considered a growing environmental crisis. The authors urged a broadscale approach involving writers, artists, storytellers, film and television productions, musicians, and social media to engage in a communications blitz to bring about a shift in social consciousness about the warming climate.[34]

The Oregon Climate Change Research Institute, created by the Oregon legislature in 2007, offered a stark appraisal of global and regional warming in 2010. The "magnitude and pace of changes" were unprecedented, the institute reported: temperatures had been warmer in the last few decades than in some 120,000 years. Rising temperatures were clearly attributable to human activities, especially burning fossil fuels that released carbon dioxide and heat-trapping gases. Warming temperatures would substantially affect Oregon, the greatest harm in reduced snowpack and decreases in summer water supplies. The institute reported that, by 2050, snowpacks would be 50 percent less than in the twentieth century, posing a serious threat to agricultural production. Wildfires and coastal flooding would increase, plant species, terrestrial animals, and fresh and saltwater species would migrate to different areas. The Oregon climate report concluded that the forces driving climate change were "population, consumption, and the intensity of the economy."[35]

Environmental change and the warming climate attracted increasing attention across the forested regions of the Pacific Northwest, including fears that rising temperatures would jeopardize scientific assessments associated with the Northwest Forest Plan. Strategies developed since 1994 to promote long-term biodiversity—of species, their habitats, and the characteristics of forest environments—faced an uncertain future under changing climatic conditions. Large wildfires, Tom Spies and colleagues wrote in 2010, had already wreaked havoc with thousands of acres of old-growth forests. Competing species and diseases were threatening plant and animal species already at risk. Because existing policies were grounded in assumptions of mild climatic and disturbance regimes, they feared that "forest policies and practices may not be well-designed to deal with climate change."[36]

Scientists, according to the Spies article, had designed programs to protect old-growth forests without considering climate change. The only activities addressing a warming climate were related to reducing carbon emissions. At the same time, a majority of scientists agreed about the potential effects of sharply rising temperatures. With relatively modest climate changes in the next thirty years, there should be enough time to use recent "ecologically-based policies" to develop strategies to cope with changing temperatures. While Spies and colleagues focused on federal lands, they provided context for the entire Pacific Northwest to cope with changing conditions and to identify adaptations to constrain the rate of change.[37]

With projections limited to thirty years, Spies and associates assessed the legal framework's ability to adjust to environmental change. The authors believed the language in the Northwest Forest Plan was sufficiently flexible to provide protection for habitats and ecosystems. Although legal terminology was "broad and appears to permit adaptive actions," the wording could be revised if it was too restrictive. There was a question, however, about policymakers—whether they would agree to adjustments and assure that funding was available. Because uncertainties abounded, administrators should develop management practices that were more effective than the failed adaptive programs under the Northwest Forest Plan. Efforts to implement adaptive management, they noted, had been "ecologically and socially double-edged swords." In an age of climate change, effective policies need to be grounded in good public relations, smart legal strategies, and solid economic arguments.[38]

The Andrews Forest and the LTER Network were increasingly involved in a wide range of studies related to climate change. In April 2012, *BioScience* published a series of articles from the LTER Network reflecting the importance of long-term research for understanding future environmental change, including a warming climate. When NSF funded the initial LTER sites in 1980, it did not envision the scientific findings that long-term research would contribute in the face of rising global temperatures in the twenty-first century. The twenty-six LTER sites in 2012 represented marine, coastal, tundra, and aquatic ecosystems that had accumulated a wealth of multi-decade data on the global environment. The network cited experiments at Harvard Forest in soil-warming research and the dynamics of carbon dioxide losses from soil when it is heated to experimentally simulate global warming. The *BioScience* authors argued that the LTER Network needed to make a major effort to meet the public's expectations on how to cope with global warming.[39]

The authors' objectives in the *BioScience* articles were to assist in developing a resilient and sustainable environment and to integrate ecological research in managing natural resources. In that sense, the LTER Network was ideally suited to assist with a wide array of environmental challenges, including climate change. The twenty-six LTER sites represented the organisms and processes of important biomes, their disturbance regimes, human influences, and prospects for long-term environmental change. The network's features, representing a wide array of ecological and social expertise, positioned it to understand the effects of future environmental change. "Environmental literacy," G. Philip Robertson wrote, "is an important ongoing legacy of LTER Network science."[40]

Another proposal to cope with future environmental change, scenario studies, incorporated science, social expectations, and assumptions about important environmental conditions to develop alternative predictions about the future. Lead author Jonathan Thompson, who did his graduate work at

Oregon State University and was with Smithsonian Institution's Conservation Biological Institute, headed a team of eight scientists, including Andrews personnel Fred Swanson and Tom Spies. Thompson cited the work of the American Forest Futures Projects at Harvard Forest, the Andrews, Bonanza Creek, Coweeta, and North Temperate Lakes LTER sites, where researchers were addressing differences among forested regions "in their socioecological responses to global change." The intriguing feature of scenario studies was their ability to link the past to the future, "questions about how past change may or may not help understand future change." Scenario studies, Thompson argued, would provide a flexible approach for integrating ecological science with future, unpredictable global changes.[41]

Climate change figured in the work of several Andrews scientists, including OSU geographer Julia Jones, whose research involved hydrology and other water investigations. Jones worked with Heejun Chang, of Portland State University, studying seasonal streamflow in several western Cascades drainage basins. Declining snowpack, they discovered, negatively affected spring streamflow and contributed to low water in the late summer months. With rising temperatures, seasonal water supplies were expected to change as warmer summers increased evapotranspiration, thereby decreasing streamflow. The authors observed that future variations would be more pronounced in mountainous uplands, where snow is important, than at lower elevations. There was another component to rising atmospheric temperatures and low summer streamflows: warmer water in streams would have severe implications for salmon survival, especially in eastern Oregon drainage basins. Most studies at this time (2010), however, offered few plans to adjust to those conditions.[42]

In another nuanced publication, Jones continued her observations about streams and climate change, suggesting there was not always a straight line between streamflow and warming temperatures. In some instances, the volume of water in streams reflected past disturbances, some of them anthropogenic. Forest management practices and regrowing cutover areas ("gradual forest succession") could influence the rate and timing of snowmelt. Jones also participated with graduate student Kendra Hatcher in a major study of climate and streamflow in the Columbia Basin using hydrologic models indicating the timing of snowmelt and late-summer drought, conditions that would reduce the availability of water to downstream communities. Hatcher and Jones suggested that conifer forests in headwaters streams would lessen the loss of water. Those ecological conditions and the hundreds of dams on the Columbia River

The research of Julia Jones, an OSU professor of geography, involved hydrology and other water investigations, including the effects of climate change on those issues.

system would provide some "resilience to climate change." They estimated that ecological conditions in the upper basins and downstream engineering features (reservoirs) would help regulate streamflow.[43]

Under the supervision of Julia Jones, Kathleen Moore completed an OSU doctoral dissertation in geography assessing the impact of climate change on the Army Corps of Engineers thirteen reservoirs in the Willamette River Basin. Primarily for purposes of flood control, storage, and recreation, Moore argued that treating the regulatory dams as a single system would be the best strategy to cope with increased risk of flooding in winter and reduced streamflow in the summer. Compared to the historical practice of raising the pools behind the dams later in the winter, she thought it prudent to begin slowly filling the reservoirs earlier in the season. Because the reservoirs could not serve "both flood damage reduction and water storage at the same time," adjustments at the dams needed to be timed to ensure sufficient water storage for summer use. Changes should be made, Moore argued, to alter the "reservoir fill path," the timing of which was determined when the dams were constructed, most of them in the 1950s and 1960s. The looming threat of a changing climate, the increasing threat of winter floods and scarce supplies of water in the late summer, made the old formula inaccurate. Given increases in winter flooding, "the optimal reservoir fill path would shift earlier in the water year." Put simply, "reservoir management may need to adapt to future changes in water supply and demand."[44]

Three Oregon State University faculty published a more detailed and data-driven study in 2017, citing the precarious situation in the Willamette Basin, where only a very slight rise in temperatures could turn snow to rain and contribute "to ski area closures, recreation restrictions, municipal water limitations, harmful algal blooms, and high fish mortality." Using data from the extraordinarily low-snowpack seasons of 2013–2014 and 2014–2015, Eric Sproles, Travis Roth, and Anne Nolin highlighted the significance of declining snowpack ("snow drought") in the Willamette Basin. Historically, heavy winter precipitation in the western Cascades, and a lesser volume in the Coast Range, positioned the Willamette as the thirteenth largest river flowing within the United States. The authors noted that the *Andrews Forest Newsletter* reported in the fall of 2015 that streamflow in Lookout Creek had set an all-time low for the sixty years that records had been kept.[45]

The McKenzie Basin, a significant contributor to the Willamette's "at risk" snow zone, Sproles, Roth, and Nolin wrote, made up only 10 percent of the Willamette systems landscape, yet it provided 25 percent of the river's flow at its confluence with the Columbia. In the high elevations of McKenzie's headwaters, snowmelt contributed to its relatively even flow throughout the year. The looming threat for the Willamette's main stem rests in historical climatological data indicating that snow accumulates at approximately freezing temperatures, and only a slight increase would turn it to rain. For the McKenzie, changing weather patterns and warming temperatures already show that peak streamflow had shifted earlier in the year. The Natural Resources Conservation Service's Snow Telemetry (SNOTEL) Network sites traditionally provided important data; however, their location at lower elevations limits their utility, because they do not reflect conditions at higher-elevation snow zones.[46]

The warming climate and downward-trending snowpack were instructive for scientists (and historians), the authors observed, because historical climatological records are no longer useful in predicting the future. The newest figures suggest that reservoir managers needed to create new predictive models to recalibrate water storage in the Willamette Basin's thirteen reservoirs. The shift from snow to rain, especially at mid-elevations, would likely affect the recharge of groundwater, the OSU scientists estimating that it would take seven years before such data will be available for the low-flow years of 2013–2014 and 2014–2015. They advised a new approach, allowing water managers to develop strategies to adapt to a very different future.[47]

A Yale University student's undergraduate research project contributed to the differing nature of the experimental forest's research. Anna Young's senior essay, which focused on wild pollinators in the high meadows of the Lookout Creek drainage, revealed that, during a five-year period (2011–2015), the snow regularly melted earlier, with the greatest abundance of flowers blooming five weeks earlier than historic averages. Her data suggested that wild pollinators were adapting, although with significant declines in interactions between plant and pollinator. Her research, however, suggested significant problems far from the Andrews Forest, in California's Central Valley where "the threat of climate change to wild pollinators exacerbates the susceptibility of our agriculture industries to a global pollination crisis." For the long-term sustainability of pollination, and considering persisting die-offs in commercial hives, Young urged agriculturalists to diversify their pollination strategies and plant hedgerows and the margins of fields with wildflowers to encourage the proliferation of wild bees.[48]

Modest climate change policies have abounded across Oregon, the nation, and much of the globe for nearly three decades. On the Andrews, where forest scientists had been debating how best to sequester and store carbon, some raised questions about the extent to which forests should be managed to achieve carbon sequestration. Parallel with those discussions, Chelsea Batavia and Michael Nelson, graduate student and LTER principal investigator, respectively, addressed ethical issues associated with the carbon question. Since the adoption of the Northwest Forest Plan, they observed, there had been marked increases in carbon sequestration in the region's forests. Although federal land managers had developed mitigation strategies for climate change, little progress had been made in implementing carbon sequestration proposals on land-management operations. Appeals to "the public good" had failed to elicit redirections in forest polices. Managers who were following multiple-use practices, Batavia and Nelson reported, "must decide how to execute policy in a way that appropriately balances diverse interests at multiple scales," decisions that required ethical judgments about what should be represented. They noted that the problem was rooted in the human tendency to make decisions with short-term objectives, whereas ethical considerations across broad scales of time and space should be incorporated when managers were implementing climate-mitigation policies.[49]

During the second decade of the twenty-first century, in addition to studying warming temperatures, Andrews scientists continued to generate innovative management strategies for federal forest lands. Senior scientists Jerry Franklin

and K. Norman Johnson, who had partnered with Franklin on several projects, set the standard, drafting proposals for restoring national forests on the
eastern and western sides of the Cascade Range in Oregon and Washington.
Discussing both the moist west side and the dry east side forests (with some
sections of southwestern Oregon in the latter category), Franklin and Johnson
wanted to enhance the ecological and economic value of the forests through
reserving older trees, thinning stands to promote structural complexity, and
harvesting younger stands to promote diverse age classes and species of trees.
Because species composition and precipitation patterns differed east and
west of the Cascades, they proposed silvicultural prescriptions to align with
environmental conditions. Published in the *Journal of Forestry* in 2012 (with
accompanying articles about pilot projects), the authors advanced ideas to
promote "ecological integrity and resilience" on federal forest lands.[50]

Because ecological restoration had become the standard for managing federal lands, Franklin and Johnson presented a solution that recognized the new
"sociopolitical reality." Grounded in ecosystem science, they acknowledged
that ecological restorations required commercial timber harvests if they were
to be widely implemented. Furthermore, such an effort should be ecosystem-
wide rather than focusing on single phenomena. National forest restoration
should focus on functional and resilient forests, beginning with degraded environments where human activities had introduced major disturbances. Equally
important, managers should avoid a "one-size-fits-all approach," because ecosystems differed east and west of the Cascades. The authors proposed detailed
prescriptions for the moist west and dry east forests. Working with stakeholders in pilot projects on national forests would improve the public's trust in
the Forest Service and BLM. In the eastern dry forests, where catastrophic
wildfires posed severe threats, a twenty-year plan for restoring some 50 percent of the forested landscape outside of wilderness and roadless areas would
involve increased commercial thinning. For the moist side of the Cascades,
regeneration harvests would provide a long-term yield of timber that presently
did not exist.[51]

Abe Wheeler headed up the "Roseburg Pilot Project" of 450 acres on
BLM land in the Myrtle Creek watershed. The requirements that led to its
selection included the age of trees, habitat characteristics, its location within
the range of the northern spotted owl, and the regrowth history of the area.
The silvicultural design would "protect or enhance ecological values." The
Roseburg Pilot Project was not a cookie-cutter model for forest management
but involved restoration strategies based on "using what the stands had to

offer." Wheeler praised the design for offering ecological benefits and providing wood fiber. The Pilot Joe Project, the dry restoration proposal, was on an 80,000-acre landscape in the Middle Applegate watershed where cool, moist winters and hot, dry summers prevailed. The area included dry, south-facing slopes of hardwoods and shrubs that were historically subject to frequent wildfire. Douglas-fir, ponderosa pine, black oak, and Pacific madrone dominated the moist-facing slopes. The restoration effort involved shifting the species composition on the slopes from Douglas-fir to other species to improve "survivability" of trees older than 150 years. Ed Reilly, project coordinator for Pilot Joe, judged the project a success, because it provided jobs, lessened forest density, and reduced the fire danger.

Among the many foreign scholars participating in collaborative research projects with the Andrews, graduate students from overseas universities have made use of historical information on the Andrews website to craft their graduate theses. Katerina Honzakova of the University of Bayreuth in Germany used publicly accessible data to determine the effects of climate change at various elevations in the western Cascades. Employing numbers from the weather station at the Andrews headquarters and six SNOTEL Network sites in the western Cascades, she examined temperature patterns from valley and ridge stations. Readings from valley locations, she discovered, showed noticeable warming trends, the most pronounced being the earlier date of last frost in the spring. Her investigation revealed little difference in the length of the vegetation growth period at most stations. Andrews forest director, Mark Schulze, indicated that Honzakova, who never visited the Andrews, was part of a trend over the last few years of students basing their work "entirely on data generated by others." It was appropriate, he believed, that publicly available information made this possible, but "feels uncomfortable with student projects that don't involve any field time."[52]

Technological advances through the years have enabled scientists to examine the aftereffects of timber harvests on the Andrews Experimental Forest between 1950 and the mid-1980s, just before the cessation of conventional national forest timber sales under the Northwest Forest Plan. When David Bell, Tom Spies, and Robert Pabst examined the effects of the early harvests decades later, they had access to great advances in lidar (light detection and range) remote sensing, the ability to use light-detecting sensors to measure en masse—tree height, diameter, density, and other characteristics—across large areas. Their 2017 study focused on edge environments and how clear-cut harvests influenced the ecological reach into old-growth timber. "Clearcutting," they learned, "can alter the understory microenvironment in a neighboring old-growth forest at distances of tens to hundreds of meters." The edge of harvests affected the growth of trees and the density of dead trees.[53]

The development of high-resolution remote sensing also provided the ability to measure the horizontal and vertical characteristics of vegetation, advancing the ability to characterize forest stands. Bell, Spies, and Pabst observed that the new technology could portray the collective basal area (a cross section of tree measurements at diameter breast height), biomass, density, individual trees, and wildlife habitat. For their purposes, the value of lidar was its ability "to directly measure the magnitude and scale of influence on forest structure across large areas." They used lidar to assess the edge influence of clear-cuts on the Andrews, finding that, years later, uncut landscapes had reduced basal areas as far as eighty yards from the edge of harvests. The extent of those influences increased with elevation. Greater mortality and declining basal area were also associated with exposure to high winds and storms.[54]

Another study of harvests on the Andrews and nearby Willamette Forest streams focused on the corpus of aquatic life following clear-cuts. Matthew Kaylor and Dana Warren compared data from earlier studies on five headwater streams in the immediate aftermath of harvests and four decades later, when the regrowing forest developed canopies to shade creeks. The scientists' data indicated that harvesting riparian trees dramatically increased exposure to light, amplifying the availability of food for invertebrates and fish, principally cutthroat trout. Rising temperatures also played into this scenario, a problem for downstream salmonids habituated to colder water. With the passing of time and the closure of canopies decades later, invertebrate life and cutthroat trout declined. "Stream light availability" over the years affected the number of cutthroat trout and other life-forms in the five streams. The authors concluded that reductions in certain life-forms can be expected to follow the closure of canopies.[55]

Beyond the experimental forest, Andrews personnel continued their participation in the Northwest Forest Plan, the wide-ranging and amended policy agenda for managing the twenty-four million acres on the seventeen national forests in Washington, Oregon, and northern California. The Pacific Northwest Research Station published a three-volume *Synthesis of Science* in 2018, covering the achievements and shortcomings of the plan since its implementation in 1994. The emphasis of the assessment—a "science synthesis"—provided a summary of important scientific advances during the previous twenty-four years. The *Synthesis* was forward-looking, an effort to present the management accomplishments on federal lands in western Oregon's national forests and BLM lands. Although the Forest Service and BLM operated under separate planning processes, the two agencies shared common objectives.[56]

Under the leadership of Pacific Northwest Station scientist Tom Spies, the assessment reviewed the Northwest Forest Plan's revised rule adopted in 2012—highlighting the importance of ecosystem-wide approaches, in lieu of the older emphasis on the viability of single species. The new rule reduced the burden of dealing with multitudes of species, of "survey and manage," or, in conservation biology jargon, shifted from intense "fine filter" to "coarse filter" management. The revision removed the requirement for searching for every species in a designated stand of timber, and simplified the management policies on national forests facing new threats from invasive species, wildfire, and climate change. Other changes were afoot as well: the emergence of amenity-based economies in some forest communities, the timber industry shifting to smaller logs, and efforts to enhance collaborative approaches to managing national forests. The major focus of the 2018 publication, however, was on the significant scientific advances since the adoption of the Northwest Forest Plan.[57]

Spies and his colleagues divided the twelve chapters of the *Synthesis* among major science-related problems that managers confronted—climate change, old-growth trees, the northern spotted owl, marbled murrelet and other species, aquatic conservation, socioeconomics and forest management, changing public values, environmental justice, tribal issues, and the integration of ecology and social science to inform managers. The lengthy assessments of climate change were critically important, because the Northwest Forest Plan had skirted the issue in 1994. In the decades since the adoption of the plan, climate change had become "an overarching theme in natural resource science and management." What could be expected in the future were warmer and drier summers and warmer and wetter winters.[58]

The exhaustive review of the Northwest Forest Plan in the *Synthesis* praised its science-based strategy to protect old-growth forests and the species dependent on those habitats. The study pointed to the obvious, however: the plan failed to generate the anticipated levels of timber production. Although the Forest Service had made strides in improving its understanding of the relations between social and ecological systems, declining federal budgets, lack of markets for smaller logs, and closures of sawmills limited the agency's ability to achieve its goals of ecological restoration. At the same time, the Forest Service was aware that some communities near national forests were thriving through recreation and amenity-based activities, while others needed access to timber harvests to survive.[59]

Spies closed the *Synthesis* assessment with an appeal to the Forest Service to revise its strategy to accommodate the barred owl's threat to the northern spotted owl, the increasing dangers of wildfire, and the declining forage and species diversity in dry forested regions. In addition, a "declining forest infrastructure"—the lack of workers and mills to enable the agency to actively manage forests to improve resiliency—limited the Forest Service's ability to make jobs available in many communities. The larger effort to meet "old and new agency goals" remained elusive. The problems rested in continued weaknesses in Forest Service/stakeholder collaboration, and the failure to convince communities to agree to experimental risks to achieve ecologically sound forests and to provide supplies of wood fiber.[60]

A less-heralded venture at the Andrews has been its arts and humanities programs, which play artfully on scientific achievements at the experimental forest to portray the wonders of nature through creative writing, poetry, fiber artistry, visual and musical productions, and other inspiring work. The Long-Term Ecological Reflections program, a place-based strategy for integrating the sciences and arts and humanities, explores the connections between the ecological and the human spirit to craft essays and poetry, photography, painting, and myriad other art forms. The emphasis of the program, initiated at the Andrews, with comparable activities at other LTER sites, is to take the long view of ecological change, whether natural or human induced, to portray symbolic references to the physical world, in this case, the forested landscape of the Lookout Creek drainage. The Reflections program, now closing on nearly two decades of productive imaginative work, has become an important component of activity on the experimental forest. Beyond the Andrews, ecological reflections programs range across New England's hardwood forests to the Florida Everglades, Wisconsin's northern temperate lakes, and central Alaska.

Chapter Seven
Long-Term Ecological Reflections

What links the characters is survival—the survival of both trees and human beings. The bulk of the action unfolds during the timber wars of the late 1990s, as the characters coalesce on the Pacific coast to save old-growth sequoia from logging concerns.

—*Kirkus Reviews*[1]

The sciences and humanities have had a living presence on the H. J. Andrews Experimental Forest since the early twenty-first century. Blending the ideas of Jim Sedell, Kathleen Dean Moore, and Fred Swanson, the Long-Term Ecological Reflections program reflects a long historical tradition in the United States, beginning with federally funded continental railroad surveys in the 1850s and the great exploring expeditions of the Corps of Topographical Engineers after the Civil War. Their published reports, photography, sketching, and paintings formed a corpus of knowledge illustrating the early ties between science and public policy. The railroad surveys provided the first reputable scientific knowledge about the American West, its rivers, valleys, mountains, and awe-inspiring basins and plateaus. The four post–Civil War surveys—by Clarence King, Ferdinand V. Hayden, George M. Wheeler, and John Wesley Powell—produced maps, epic paintings, and stunning photographs. After the National Park Service was created in 1916, it exploited the spectacular scenery in the parks to attract visitors. In subsequent decades the Park Service began sponsoring artist residencies and displaying their work at visitor centers. More recently, the National Science Foundation joined that tradition, sponsoring the Antarctic Artists and Writers Program since the 1950s. The Organization of Biological Field Stations and the National Association of Marine Laboratories also have taken advantage of their locales to establish arts and humanities agendas.[2]

In the Pacific Northwest, the Endangered Species Act, the northern spotted owl decisions, and the ongoing struggle over old-growth and ancient forests spurred debates about the meaning and value of the federal estate, events that contributed to a profusion of books and articles extolling the beauty and majesty of the region's spectacular landscapes. Writers, landscape photographers, and wilderness advocates produced an impressive volume of print material from the 1980s into the present century. Robert Michael "Bob" Pyle was among them, winning the John Burroughs Medal for Nature Writing for *Wintergreen: Rambles in a Ravaged Land* (1986). Pyle's ruminations involved the Willapa Hills in Washington's Grays River backcountry, where he makes his home. His affectionate tale describes a century of logging and its negative effects on people, the land, and plants and animals. The well-traveled Pyle has lectured widely, even globally, on natural history, conducted writing workshops at colleges and universities, and delivered numerous keynote addresses.[3]

The *Seattle Times'* Pulitzer Prize–winning science writer, journalist William Dietrich, published *The Final Forest* in 1992, a riveting story of the polarizing controversy over logging vast tracts of old-growth timber on Washington's Olympic Peninsula. Focusing on the logging town of Forks, Dietrich interviewed loggers, truck drivers, timber industry executives, foresters, environmentalists, and regional politicians, inquiring into their opposing social and economic views. Loggers on the Olympic Peninsula, the author argues, were caught between timber corporations and environmentalists and the federal government's misguided policies of clear-cutting thousands of acres of old-growth trees. Dietrich reprised Eric Forsman's research linking the northern spotted owl with old-growth forests, noting that "the battle lines were drawn." *Kirkus Reviews* praised *Final Forest* for providing a "poetic description of the battleground, a moving assessment of an ecological dispute with global implications."[4] It is important to recognize that Pyle, Dietrich, and other writers were giving a living presence to the intersection between landscapes and human perceptions of them.

Far from the Pacific Northwest, Alison Deming published *Science and Other Poems* (1994) through the Louisiana State University Press, a collection applauding collaboration between scientists and creative writers. The book won the Whitman Award of the Academy of American Poets, with the academy's Gerald Stern praising her precision of language, brilliant metaphors, and clear-eyed observations and wisdom. Deming lives in this world, he observed, and "it is the world she settles for . . . even if she must rage a little." The LSU Press extolled her artistic approach to scientific inquiry, using her poet's touch for spinning out parallels between the physical world and "the dark mysteries of the human heart."[5] Deming and Robert Pyle, among the first

Kathleen Dean Moore, OSU Emeritus Distinguished Professor of Philosophy, founded the Spring Creek Project for Ideas, Nature, and the Written Word, which has partnered with the Andrews Long-Term Ecological Research program to form the LTER Reflections program, a humanities-arts-science collaboration.

writers-in-residence at the H. J. Andrews Experimental Forest, would set a high bar for those who followed.

The Andrews Forest's writing/arts program was a collaborative effort with the Spring Creek Project for Ideas, Nature, and the Written Word, a privately endowed enterprise in Oregon State University's Department of Philosophy. Nature writer Kathleen Dean Moore established Spring Creek, which now operates out of the School of History, Philosophy, and Religious Studies. Spring Creek has hosted writers and artists at a mountain cabin in the Coast Range and partnered with the Andrews Forest LTER to create the Long-Term Ecological Reflections program. Spring Creek offers residencies and has held symposia at the Andrews, Mount St. Helens, and other locations. It has been a valued partner in the experimental forest's arts and humanities initiative.[6]

The arts and humanities programs at the Andrews grew out of the National Science Foundation designating the initial six Long-Term Ecological Research (LTER) sites in 1980, one of them the Andrews Forest. Fred Swanson, an originator of the Andrews Forest Long-Term Ecological Reflections program, pays tribute to Jim Sedell, an aquatic ecologist with the Pacific Northwest Research Station, and Kathleen Dean Moore, who had published *Riverwalking: Reflections on Moving Water* in 1996. After reading *Riverwalking*, Sedell invited Moore in 2002 to lead a workshop, New Metaphors for Restoration, at the Andrews Forest and to participate in a public event at Starker Arts Park in

Corvallis. Swanson met Moore in the process of transferring funds for the events. Those exchanges led to Spring Creek hosting the first writers in residency at the Andrews in 2004. Since then Spring Creek and Moore, now retired, have been important contributors to the Reflections program.[7]

Although the Andrews Experimental Forest was the first of the LTER sites to offer a writers/artists residency program, Harvard Forest in Massachusetts has been a pioneer in such initiatives, drawing on a wealth of eclectic scientific sources involving some three centuries of changing landscapes. The forest's arts and humanities creative work can be traced to the forest's founding in 1907, and writer and artist Richard Fisher. During the late 1930s, under director Al Cline's supervision, Harvard staff constructed intricate scale-model dioramas portraying the successive stages of land use and forest succession. The eight delightful dioramas are on public display at Harvard Forest's Fisher Museum in Petersham, Massachusetts. To those who knew and appreciated Fisher, the dioramas were clear representations of his appreciation for landscape history, art, and storytelling.[8]

Central to some of those observations were Henry David Thoreau's descriptions of mid-nineteenth-century Massachusetts. David Foster, who has enjoyed a long tenure as director at Harvard Forest, recognized the timeless qualities of humans and their association with landscapes when he was building a cabin in Vermont in 1977. Fond of reading Thoreau's journals, Foster learned that the countryside Thoreau traveled through between 1837 and 1861 was much different from the thick New England forests of the late twentieth century. In reality, Thoreau's journeys took place across meadows and farmland, interposed with occasional woodlands. Foster subsequently published *Thoreau's Country: Journey through a Transformed Landscape* (1999), a portrait confirming that Thoreau's landscapes had changed dramatically through the passing decades. From Foster's book, readers learn that natural and human disturbances have always affected forest ecosystems.[9] Those insights would be reflected in the writers and artists renditions of the experiences on the Andrews Forest.

LTER's *Network News* featured the Andrews Long-Term Ecological Reflections program in April 2005, describing the operation as scientists "enjoying a new collaboration with creative writers." Partnering with the Spring Creek Project, the initiative supported writers and humanists interested in exploring changing human relationships with nature over long periods of time. *Network News* quoted Spring Creek's Kathy Moore, who described their cooperative enterprise as an effort to bridge the sciences and humanities to help people live sustainably in a world "threatened by cascading changes." The *News* mentioned Robert Michael Pyle's essay, "The Long View," and his visit to scientist Mark Harmon's log-decomposition site. "The long view," Pyle wrote, "requires faith in the future, even if you won't be there to see it for yourself." *Network News* informed readers that such work would be posted on the Andrews website under the rubric, The Forest Log.[10]

With the Reflections program in its sixth year, Fred Swanson, Charles Goodrich, and Kathleen Dean Moore emphasized the importance of the Andrews Forest offering both scientists and humanists compelling landscapes to stir scientific inquiry and inspire creative work. The authors believed the humanities offered counsel to Forest Service administrators in their relations with the public and managing federal lands. Although Reflections was still a relatively new enterprise in 2008, the authors contended that it was drawing attention to the creative strength "that flows from the conversations between art and science, and between writers and scientists." The Andrews Reflections program had already influenced the establishment of arts and science initiatives at North Temperate Lakes LTER (Wisconsin), Bonanza Creek LTER (Alaska), and at the Shaver's Creek Environmental Center at Pennsylvania State University.[11]

Writing for *Science Findings* in 2008, a Pacific Northwest Research Station publication, Corvallis-based ecologist and science journalist Jonathan Thompson praised the Reflections program for bringing "together scientists, creative writers, and environmental philosophers to consider ways to conceptualize and communicate views of long-term ecological change in forests and watersheds and the participation of humans in that change." The program would integrate humanities with conventional science "to learn about the special places for which the Forest Service is responsible." Thompson applauded the Reflections program for bringing creative writers and philosophers to the Andrews and Mount St. Helens to engage in ecological inquiry while associating with scientists. Spring Creek's Charles Goodrich told Thompson that the program received twenty to thirty applications a year, with inquiries coming from across the United States. The prize for invitees was minuscule—"to spend a week at the Andrews" free of charge. From their apartments, the visitors were asked to visit three sites where long-term ecological research had been taking place for decades.[12]

A decade and more after the launch of the Andrews Reflections program, science/ecology/humanities interactions were convening at twenty of twenty-five LTER sites, some forty biological field stations and marine laboratories, and approximately fifty National Park Service locations (the latter as writer/ artist residencies). Although there were vast differences among the sites— from the South Pole to the City of Baltimore and to Bonanza Creek in Alaska—place was critically important to participating artists and writers. Fred Swanson believed that relations between artists/writers and ecologists varied across environmental/institutional locales. Although the physical settings differed significantly, the common bond for resident writers and artists was communication with the public.[13] Swanson identified the critical relation between the presence of humanities and science personnel on the Andrews in the midst of a place with magnificent topography, still stupendous old-growth forests, and a profusion of animal life—all of it fodder for inquiry.

Swanson reported in 2015 that the Andrews Forest had hosted some fifty writers and artists whose work appeared in prominent literary journals such as *Orion* and *The Atlantic*. Robin Kimmerer, a botanist, writer, and distinguished professor in the College of Environmental Science and Forestry at the State University of New York in Syracuse, was a Blue River Fellow at the Andrews in 2004, where she contributed three essays, "Interview with a Watershed," "Listening to Water," and a published piece, "Witness to the Rain," in Thomas Lowe Fleischner's *The Way of Natural History*. Kimmerer's published piece is steeped in the midst of western Oregon's rainy season, where land and water appear as one: "Here in these misty forests those edges seem to blur: rain so fine and constant as to be indistinguishable from air, cedars wrapped with cloud so dense that only their outline forms emerge."

In "Listening to Water," Kimmerer addresses the essence of long-term reflections:

> I feel my number acutely. Two of two hundred, the second writer to be summoned to this place. The second voice in a chronicle intended to stretch out for 200 years. Two hundred years is young for the trees whose tops this morning are hung with mist. It's an eye blink of time for the river that I hear through my open window and nothing at all for the rocks.[14]

In another essay, "Interview with a Watershed," Kimmerer visits the three small watersheds where experiments and data collection began in the 1950s. She relates their story, one watershed left uncut, a second clear-cut, and a third partially cut and the logs removed via a skyline crane. Gauging stations at the bottom of each watershed recorded data on the volume of water flowing and its chemical makeup. The statistical information from the clear-cut watershed revealed "a landscape hemorrhaging nutrients and filling the stream pools with sediment as the soil washed away." In the absence of shade, the stream's temperature was too warm for fish. In the unharvested watershed, the old-growth forest shaded the stream, with water running cold and clear. Kimmerer concluded, "Water is a storyteller," reading and learning from water has been pivotal in reshaping how we manage forests. And, more personally, "doing the field work" gave her a sense of being "intimate with the place" and provided a better understanding of the science.[15] Fieldwork, the domain of scientists, was also the province of the humanities.

Alison Deming, a Blue River Fellow in 2006, posted several poems and essays from her experience at the Andrews. In "Attending to the Beautiful Mess of the World," she tells an amusing story about the visit of fifth- and sixth-grade students, bored listening to scientists explaining weather-related information, becoming enthralled when they see newts and salamanders. The substance of the essay, however, was her participation in the Long-Term Ecological Reflections program, "a concomitant to scientific work being done under the umbrella of Long-Term Ecological Research." Deming visited the forest's well-known study of decaying logs, "morticulture," investigating the dead. She traveled with Fred Swanson on Forest Road 130, "expecting science to be the boss here," because her host had been studying the place for decades. She soon realized, however, that Swanson was treating her as "a partner in the enterprise of understanding this place." Deming discovered what other visitors to the Andrews learn: "Fred has an excited mind and verbal acuity that are hard to keep up with, especially when his commentary is filled with a lifetime of learning about this forest."[16]

In a brief essay, Deming described fieldwork with Steve Ackers, who had been studying northern spotted owls for seven years. She accompanied Ackers to Hardy Ridge on the South Fork of the McKenzie River, climbing to a site "better suited to flying squirrels and red-backed voles." When Ackers located an owl, the fascinated Deming reported that all the literature and data on the owl cannot compare with seeing one in the wild—the spotted breast, barred tail, and a face spangled with brown semicircles. The configuration of colors and lines making her eyes seem the size of her head, "The blackness of her pupils is so pure they look like portals into the universe." Deming added that she made the trek to the owl's location for beauty, "a beauty enhanced, not diminished by science."[17]

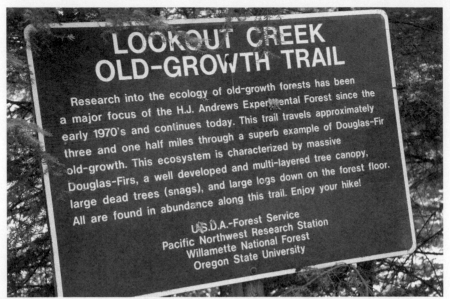

Lookout Creek Old-Growth Trail passes three and a half miles through massive old-growth
Douglas-fir. Hundreds of visitors have walked this trail through the decades.

Kathie Durbin, a veteran environmental journalist who wrote for the
Eugene *Register-Guard*, *Willamette Week*, the *Oregonian*, and Vancouver,
Washington's *Columbian*, visited the Andrews in 2007. An award-winning
environmental writer, she has also authored three books on her beloved
Pacific Northwest. Durbin's essay, "Reflections on Change, Natural and
Otherwise," provides substance to her long-standing observations about dis-
tressed and natural forest environments across the greater Northwest. Sitting
along Lookout Creek, she writes as a person thoroughly versed in Northwest
plant life—"too late for trilliums . . . too early for rhododendron." Durbin
noticed fresh growth everywhere: Oregon grape's "glossy dark serrated leaves,"
shiny new poison oak "with its shapely oak clusters," and vine maple's "bright
mint-green leaves." In a series of daily journal entries, she tells about travers-
ing Lookout Creek's Old Growth Trail, a clear-cut, the log-decomposition
site, and the end of Road 320, "Jerry's Forest." She described the latter as "an
old battlefield," where Franklin toured with visitors to show them his "sloppy
clearcuts" (following the dictates of new forestry). Years later, visiting one of
Franklin's new forestry experiments, Durbin described the scene as "elegant,"
aesthetically pleasing, "a gentle, managed landscape."[18]

The month before Durbin visited the Andrews Forest, her fifty-seven-
year-old brother died in his sleep. She reflected on his passing as she sat along
Lookout Creek below a decaying log, realizing that in death the log supported

"a gallery for insects, a nursery for tiny hemlock seedlings, a banquet for fungi." She pondered whether, if her body lay beside the log for two centuries, "could I pinpoint the moment at which it returns to soil?" In fifty-seven years, a Douglas-fir would be an adolescent, a mere wink of an eye in the life of a forest. For humans, fifty-seven years would bring one to maturity, "time to reap the rewards of a life well and fully lived." In the Andrews Forest, Durbin reflected, time heals wounds "inflicted by nature and man." Relaxing by the rushing creek, she saluted the natural world for "healing my grieving spirit."[19]

Bill Sherwonit, a nature writer from Anchorage, Alaska, who trekked around the Lookout drainage in October 2006, was especially impressed with the morning wake-up call of varied thrushes. Of the twenty-five birds he identified during his stay, he thought the thrush's song, with its "haunting series of repeated one-note trills," was the most interesting and "among the most distinctive avian woodland voices." In his mid-forties, Sherwonit acknowledged he was a latecomer to appreciating nature: he began his professional career as a geologist, migrated to sports writing, outdoors newspaper reporting, and then nature writing. He left his academic field because he lacked confidence in reading "landscape's geologic history," confessing to "a love-hate relationship with science." Writing amid the experimental forest and its rich history of scientific investigations, Sherwonit believed there was "an arrogance to the culture of science [that] frustrates and unsettles me." Yet he treasured the week on the Andrews and praised its leadership for believing that creative people had something to contribute along with the scientific community.[20]

After his grumblings about scientists and the assuredness with which they went about their craft, Sherwonit praised "the work being done here" at the Andrews. What attracted his attention the most was "the forest itself: the still air, the 'natural quiet,' the softness of the forest floor and the lush carpets of moss that seem to cover everything." The forest and its old-growth stands offered a perspective on the meaning of life, a sense of eternity, of an endless universe, ever-larger environments that humble mere humans. Sherwonit walked along lower Lookout Creek and found a natural Douglas-fir footbridge with mountains of woody debris piled on top of the downed tree. Along the stream's placid pools on a quiet afternoon, he wondered "what sort of raging current could have tossed such logs about?" This jumbled mess lay "within a few feet of tranquil ponded waters," the place he chose as his personal reflection site.[21]

One of Sherwonit's informants was Tim Fox, a former archaeological crew leader on the Andrews, a surveyor, and an owl researcher. A gifted writer with publications in *Orion* and *Yes* magazines, Fox served as a Reflections Fellow in 2009. In "Primordial Chords," Fox wrote about hiking to the headwaters of Lookout Creek, where he set up camp for an overnight stay, including preparing a small fire. For one growing up in America's suburbs, he noted, "heat hidden away in wood strikes a primordial chord." With fire's embers cooled, he crawled into his tent and drifted off to sleep. Stuffing his gear into his backpack the next morning, he headed up the forest road amid darkening clouds and rain. At this point, the reader realizes that Fox is traveling during the fall season when he sees vibrant "gold and red maple leaves" glistening through their wetness. He thought about the restrictions in a designated research area; he cannot fish (unless for research purposes), and he cannot kill a ruffed grouse. Unwilling to break regulations, he continued onward and upward to the next campsite.[22]

After many years of field research on northern spotted owl demography, Tim Fox had developed a special attachment for the bird, an empathy that he associated with the owl's affinity for old-growth timber. The barred owl, a major threat to the spotted owl, began arriving in the Cascades in the mid-1960s, crossing the continent from east to west through Canada and then turning south through the Cascade and Coast Ranges. The adaptiveness of barred owls to old-growth forests "at the apparent expense of the spotted, made my heart hurt," he wrote. In February 2009, he had a transformative experience while he was walking near his home through old growth and heard the distant call of an owl. Unable to specifically identify the species, he remembered that the "interplay of tone and towering trees gave the impression that the forest itself was calling. I assumed a spotted owl, stopped and listened. The forest called again. With the voice of a barred owl." In that instant, Fox changed his views about barred owls. If old-growth systems endured—"even if the spotted owl gives way to the barred"— "the forest will retemper the barred in form and spirit to fit the mood of the trees."[23]

A quarter of a century after the Mount St. Helens eruption, in 2005, twenty-five writers, artists, and scientists rendezvoused near the volcano, where they spent four days observing, discussing, and sharing ideas about this mother of all disturbances and the rebirth of flora and fauna on the mountain. "Harnessing the power of this place," Fred Swanson wrote, "the writers and scientists reviewed the principles of terrific disturbance and renewal in all its forms—geological, ecological, and human—and what they should take away from this place." Three years following the campout on Mount St. Helens, Oregon State University Press published *In the Blast Zone: Catastrophe and Renewal on Mount St. Helens*, with Charles Goodrich, Kathleen Dean Moore, and Fred Swanson editing the slim volume. Among contributing scientists were Jerry Franklin, James Sedell, and Fred Swanson. John Daniel, Ursula K. Le Guin, Kathleen Dean Moore, Gary Snyder, and Kim Stafford were prominent writers represented.[24]

Ursula K. Le Guin, Oregon's preeminent literary figure, wrote a chapter titled "Coming Back to the Lady," about the mountain she called "my neighbor for more than forty-five years." Although her home's position in Portland's West Hills blocks a view of the setting sun, Le Guin "could see the light of the sunset on the volcano." For several months after the eruption, she and painter friend Henk Pander and photographer Ron Cronin attempted to gain permission to enter the Red Zone, a request finally granted in October 1981, when they received a one-day pass. They drove from Cougar, Washington, toward the mountain through an "endless green vitality" until they turned a corner on the mountain and entered "a world of grey ash, burnt stumps, and silence," literally passing from "flourishing life" to death. Le Guin did not return to the mountain until twenty-five years later when she accompanied the writers, artists, and scientists into the blast zone.[25]

When the eclectic group of visitors camped and hiked on the mountain in July 2005, Mount St. Helens was much different, a landscape that differed from Le Guin's memories of "snags, burnt dead trees lying in strange patterns, toothpicks, blown down by the blast." On this return visit, the once "silent dead landscape" was a healthy green and very much alive, and, as she learned from scientists, "had always been alive." Fortunately, she admitted, her science colleagues relieved her "cognitive dissonance." They also acknowledged, however, their error in predicting a slow recovery—"they rejoice in being wrong." Although she found the group friendly and interesting, her only regret was not having "silent time to sit and think and write or draw."[26]

On her trek to the campout on Mount St. Helens, Kathleen Dean Moore, an astute observer of sounds, remembered a frog's song, a winter wren chattering, a meadow lark singing, chickadees flickering among the branches, and a thunderstorm one night where she witnessed "a bolt of lightning shoot[ing] from the crater almost to the moon." Despite such comforting memories, her reflections offer a more somber assessment, sensing the force of the blast, its searing destruction as it "charged down the hillside, burning the slope to bedrock." She described the mountain as a place "where hope rides the back of horror," where the eruption "seared the eye of a robin who must have looked over her shoulder in sudden alarm." All the birds died, all the large mammals, and fifty-seven humans. There were also lessons amid this awfulness: abundant fireweed, the return of frogs, coyotes, beaver, elk, all of which prompted Moore to conclude that "life itself is a powerful force that will not be turned away." Admitting to confusion about "the difference between destruction and creation," she noted the scientists' excitement about the obvious ecological restoration taking place, what they referred to as a "great biomass disturbance event."[27]

For Kathy Moore, the trip to Mount St. Helens was a journey into the meaning of life, humans, time, and the physical environment. In her time on the mountain, she thought that she would learn acceptance, comfort, and grace. In a steep climb from the Donnybrook Viewpoint to Windy Ridge, she realized "how different I am from the mountain." Tumbling rocks along the ridge and below were reminders that human lives were not the measure of time, undercutting the belief that humans were the center of creation, that their ambitions and worries were "a special concern of the Earth's," or that humans were in control of large-scale events. "*Ha*. The mountain laughs. *Ha*, a great explosion of ash and steam, jolted by lightning."[28]

Charles Goodrich, an avid gardener, witnessed a profusion of flowers during his trekking about on the mountain, flowers everywhere, and a great variety of shrubs. The fluorescent displays were euphoric, setting his "blood to humming." Yet, the wild profusion of flowers and plants "puts the gardener in his place, aware of his limits." Climbing up a trail one day, they passed across a ridge into a spectacularly different landscape: Mount St. Helens, its crater spewing steam, a terrain of pumice, the detritus of huge landslides and rivers of pyroclastic material. Goodrich observed "crenellated ridges of seared rock and barren crags, an immense and seemingly empty landscape." There was still more to see. Even on the Pumice Plain, an area once barren of any sign of life, he saw lupine and other wildflowers and willows and red alder thriving where water was present. He also noticed different forms of plants marking the monument's boundary. The Gifford Pinchot National Forest lands were green with young hemlock and fir, whereas inside the monument (inside the national forest), plant life was greening up naturally, and more slowly. After witnessing this great disturbance twenty-five years on, Goodrich hoped the experience would help us give better care "to the refugia that our homes, gardens, and cities actually are."[29]

A Portland native, Christine Colasurdo lived in her parent's cabin during summers, a short distance from Spirit Lake. She worked at Harmony Falls Lodge in 1979, a small resort accessible by boat, or three miles by trail. The eruption on May 18, 1980, preempted her plans to work at the lodge that summer or to enjoy the beautiful hikes she took through a treasured land-scape. Years passed before she returned to beloved Spirit Lake. In Colasurdo's 1997 book, *Return to Spirit Lake*, she recounts standing at the 5,000-foot level, witnessing the end of a summer rainstorm that had just passed over Mount St. Helens. The crater was visible in the distance, and farther away, snowcapped mountains, Hood, Adams, and Rainier. "After years of shouldering the loss of a vast green landscape," this was Colasurdo's first close-up glimpse of the mountain. Finding her way out of grief was difficult. Spirit Lake, for one, "bore a strange new name—'blast zone'—to define the area where the volcano's lateral explosion had swept through." Looking at the lake below, what she saw was a "giant log raft" moving from one end of the lake to the other, depending on wind direction. Before the eruption, the lake was two hundred feet lower, shrouded in evergreens. After darkness fell and stars glistened in the night sky, she wrote, "so this was the blast zone."[30]

In the Blast Zone features five poets, Tim McNulty, John Calderazzo, John Daniel, Kim Stafford, and Gary Snyder, each of whom treated varying features of the mountain.[31]

We stand at a windy alter,
The volcano is washed in morning light.

Just out of sight,
a new lava dome heaves and falls,
swells at a pace that staggers.

      —Tim McNulty

Before it blew in 1980 I had never heard of it
never seen it from an airplane or high ridge.
Only later
did I see the photographs
as though its secret heart of fire
had been blazing all that time.

      —John Calderazzo

You did it, Mount St. Helens.
All of use looked on

you stormed in solitude,
you shrugged and shook aside
what we called beautiful
as if none of us were here,
no animals, no trees
no life at all outside.
your ancient firey joy—
I admired you mountain,
but I never loved you until now.

   —John Daniel

What if—so schooled by water, stone, flower, and tree—
I chose to be a human being?
Then I would be chosen to do what only I can do.
I would be chosen to do only what gives me joy.
Friends, you may glimpse me now and then.

   —Kim Stafford

The pristine mountain
just a little battered now
the smooth dome gone
ragged crown

   —Gary Snyder

Oregon State University Press published its first children's book in 2013, *Ellie's Log*, a story centered in the vicinity of a large tree that crashes into Lookout Creek during the wind and rain of a winter storm. Ten-year-old Ellie and friend Ricky, the protagonists, explore the forest's surroundings, experience changing seasons, and discover a wide range of plants and animals. Judith Li, a retired stream ecologist with years of experience on the Andrews Experimental Forest, authored the book, and Margaret Herring's illustrations provide benchmark features for each chapter. Li wanted *Ellie's Log*, created for children between the ages of eight and twelve, to encourage children to observe nature closely and to take field notes, and her collaborator Herring thought it important "to reassure children that they too can take notes and measure things and draw pictures." Li's objective was to provide a blend of science and literature to portray the natural wonders of the forest and all that it entails. The Oregon State University Library, a cooperating partner in the enterprise, offers a learning website and teachers' guide that complements the book (ellieslog.org).[32]

Although *Ellie's Log* is not formally linked to the National Science Foundation's Ecological Reflections program, it is a close relative, listed in NSF's LTER Schoolyard Book Series, which assures readers that research scientists reviewed the content, that illustrations were appropriate and accurate, and that the narrative should "bring ecological science to life and engage children." NSF cited the Long-Term Ecological Research (LTER) Network for providing the science, research, and long-term data for studying environmental change. The objective of the Schoolyard Book Series, therefore, is to encourage children, their families, and their communities to understand something about the Earth's ecosystems. *Ellie's Log* is one of about a dozen children's books listed on the Schoolyard website. The others represented far-flung environmental locations and topics—Colorado's San Juan Mountains, the Florida Everglades, ocean environments, a Southeastern US salt marsh, an autumn bison calf on the Great Plains, a lost seal, and a second Li/Herring creation, *Ricky's Atlas: Mapping a Land on Fire*, a sequel to *Ellie's Log*.[33]

Jim Sedell, stream ecologist extraordinaire and instrumental in the creation of the Andrews Reflections program, was fond of asking his research cohorts, "What's the story?" Sedell set in motion the invitations to creative writers and artists to spend a week or two on the forest thinking and reflecting about the place and creating cultural information regarding their experiences. The University of Washington Press published *Forest Under Story: Creative Inquiry in an Old-Growth Forest* (2016), a disparate collection of essays, field observations, and poetry from the first twelve years of the Reflections program. Editors Nathaniel Brodie, Charles Goodrich, and Fred Swanson organized the book into three parts: "Research and Revelation," "Change and Community," and "Borrowing Others' Eyes." Interspersed within each section are brief descriptive essays, "Groundwork," outlining the environmental and science/ research-related aspects of the forest.[34]

The first of the eight Groundwork essays, "Natural History of the Andrews Forest Landscape," offers a brief but incisive description of the Lookout Creek watershed, its varying elevations, geological history, seasonal weather patterns, dominant conifers and their distribution, and incidents of fire and logging, as well as the functions of streams, and an emphasis on changes that are ubiquitous across the landscape. "Decomposition" explains Mark Harmon's work as the forest's "head rotter," portraying processes of decaying logs. "Old Growth" reviews the transition in human assessments of old trees from board feet of lumber to a rich natural tapestry of ecological complexity. "Disturbance" provides a critical component of Andrews research, featuring fires, floods, and other disruptions as natural facets of changing landscapes. "Northern Spotted Owl," "Forest Practices," "Water," and "Soundscape" follow, offering brief explanations of each topic. The final Groundwork essay explains recent avian research involving sixteen listening devices and 188 listening stations across the forest. The investigation's purpose is to assess the distribution of birds and their seasonal migration patterns, the varying sounds of streams, and animal noises.[35]

The entries in *Forest Under Story* are gleaned from those in The Forest Log, on the Andrews Forest website. Some authors have multiple entries, among them Robert Michael Pyle, Alison Deming, Robin Kimmerer, and Thomas Fleischner. Freeman House, author of the award-winning *Totem Salmon* (2000) and with a long history of working on watershed restoration in northern California, was a Blue River Fellow in 2008. Reflecting on his personal experiences as a founder of the Mattole Salmon Watershed Group, House supported the concept of "communities of place," where humans assume responsibility for repairing the environmental harm they and their forebearers have inflicted on the land. Science, according to House, has always played a significant role in local stewardship, just as the Andrews Forest has defended old-growth forests. "A world without the H. J. Andrews Forest and the Long-Term Ecological Research project," he wrote, "would be a poorer place." Research on the Lookout watershed, and the scientists who worked there, made the Andrews a special place. "Their work has changed their behavior," House wrote, referring to scientists and those who join them in redemptive efforts as "science tribes."[36]

A native-born Oregonian, Laird Christensen, spent time on the Andrews in the spring of 2008 as part of a sabbatical leave from Green Mountain College in Vermont. Growing up in Oregon's Coast Range in the 1960s, he was proud of his mother's Laird family ancestors who arrived in the coastal timber country, "pulled out their axes, and began clearing the land." When he returned to Oregon in the 1990s after several years away, a different narrative was circulating among people who "preferred their forests upright to felled, bucked, and milled." Although those opposing stories persist to the present, the former hero, the timber faller, says Christensen, "now looks more like a villain." By the 1990s, clear-cuts, once symbols of progress and jobs and

timber revenue to support schools and paved roads, were equated with mind-
less resource extraction. In timber communities, the northern spotted owl was
featured in bumper stickers—SAVE A LOGGER; EAT AN OWL.[37]

As a child growing up in a logging town, clear-cuts were simply "logging,"
Christensen remembers. At age nineteen, he graded lumber in a sawmill, with
no questions about logging practices. When he returned to Oregon more than
a decade later, he viewed engineered forest landscapes through a different lens.
And then, on to college in New England, where his "interest in wild places led
him "from Thoreau to Gary Snyder, Ed Abbey, to Dave Foreman." By the time
he took a graduate fellowship at the University of Oregon in 1994, protests
opposing old-growth harvests on federal lands had turned the region into a
battlefield. Clear-cuts, in Christensen's view, had taken on a different meaning.
"So I drummed and shouted at protests, sent my poems to the *Earth First!*
Journal, and tore flagging from the paths of future logging roads." After five
years in Oregon, he left with a PhD and landed the position at Green Mountain
College as professor of English and environmental studies.[38]

Although Christensen missed the Oregon outback, he learned to treasure
Vermont's woodlands, discovering that its original forests had been cutover
but now covered 80 percent of the state. He appreciated Vermont's "middle
landscape," the pasturelands between towns and woodlands. While preparing for his Andrews residency, he acknowledged that his views about logging
had changed. Climbing up through a steep clear-cut below Lookout Ridge,
he discovered a profusion of wildflowers (and poison oak). "Who else lives
in a clear-cut?" he wrote. "There were purple peavines and tiny pink flowers
that I thought must be collomia." He saw blue vetch, black huckleberry, and
starflowers, and learned during his stay that significant experimental manipulation had been taking place at the Andrews. Christensen reminisced about a
camping venture many years ago in a clear-cut where elk began appearing at
dusk, the moon shedding light on the animals. Then he recognized, "that is the
world they live in—and I understood, with moonlit clarity, that it's the world
we live in as well."[39]

In a parallel essay, another Oregon-born visitor to the Andrews, Frederick
H. Swanson—absent from the state for forty years except for family visits—
returned to the forest from his home in Salt Lake City in late May 2014 to
spend a week as a Reflections scholar at the Andrews Forest. A graduate of
the University of Oregon, Frederick H. had taken an undergraduate class in
geomorphology from Frederick J. Swanson in 1973, when the latter held a
postdoctoral appointment at the University of Oregon. An avid outdoorsman

who enjoyed fishing, Frederick H. was interested in seeing how forests in the upper McKenzie Basin had changed, and "I also wanted to see if I had changed." He quickly observed that the Blue River drainage was well-forested, an aesthetic that differed from his youth, when the area had been targeted for heavy harvesting. "Regrowth happens fast here," he noted, "unlike in Utah." Standing abreast the ridge between Blue River and Lookout Creek on the first afternoon of his arrival, he noticed old trees, snags, and small clear-cuts on the Lookout side, and suspects that the trees were shielding logging roads.[40]

Frederick H. toured the reflection sites with Frederick J., who told him that the needs of society drove research interests on the Andrews. Frederick J. reported that, in the 1970s, scientists had learned about the effects of building logging roads; then biodiversity, the northern spotted owl, and other endangered species; and now, the focus was on climate change. Frederick H. later reflected about researchers who had spent decades on the Andrews, while he would be there for only one week. When Frederick J. told him about the "major 'runoff event'" of 1996, he described the landscape as "performing," a comment that prompted Frederick H. to write in his journal that his geomorphologist friend was "in some way delighted." Although the road at Watershed 3 had been badly damaged in the 1996 deluge, according to Frederick J., "from a purely scientific standpoint the result was fascinating." After touring through the forest, the Utah visitor confessed that he had rarely met anyone "who is both a highly accomplished scientist and a 'big picture' thinker."[41]

Frederick H. Swanson reflected about his young life of fishing and hiking along the many streams flowing into the McKenzie and worried that his nostalgic sentiments were counterproductive, because they rejected change. He recalled that the 1964 flood "blew out all my favorite fishing holes." During his stay at the Andrews Forest, he hoped "to temper [his] wistfulness." He closed his entry musing about the forest's beauty and research plots, and added a bit of irony:

> Here I am, then, a habitually impatient person, sitting on a fat log jutting into Lookout Creek, wondering if I will have anything to offer in the way of insight or understanding. I'll indulge in some nostalgic musings about life's changes in the forty years since I lived in western Oregon, and I'll observe (in a loose qualitative way) how the forest has regrown along the roads and in the clearcuts. I'll proclaim that all this is beautiful and should be protected from the money-changers. And I'll sleep well in my modern quarters, built

from 2x6s and plywood from (I would guess) the "working forest"
to the west of here, where large-scale clearcutting continues. Sweet
ironies![42]

There are several powerful essays that appear only on The Forest Log, one
of them Aaron Ellison's "Decomposition and Memory," a thoughtful narrative
on decaying logs and the human attention span. A senior research fellow in
ecology at Harvard Forest and author and coauthor of more than one hundred
scientific papers, Ellison visited the log-decomposition site in 2010 and provided a clinical description of the experiment, one that he was familiar with.
Designed to operate for two hundred years, the study would inform researchers about decomposition processes, "heretofore unappreciated by ecologists
and foresters." He welcomed the project's design—large plastic cylinders fitted into logs to measure rates of decay that, with the passing decades, might
end in an earthworm, a mushroom, or a newt. Ellison worried that the carefully designed experiment might not be adequately attended to: "Are data still
being collected from the respiration tubes; looked at and analyzed; prepared
for publication? How fast are these logs decaying?"[43]

The experimental decaying logs are impermanent, Ellison noted, forests
are impermanent, as is the Earth. He traced the cycling of carbon from the
atmosphere to growing trees and eventually to decaying wood becoming soil
and water "and back to air." But he worried about sampling intervals, interpreting data, and long-term attention to the experiment. "Will 22nd Century
graduate students be inspired to continue maintaining this experiment?"
Although decaying logs have much to tell, he feared that after twenty-five
years "the story is already fragmenting." The logs reminded Ellison of his New
England home—forests were cut, houses were built, land was bounded, and
then colonists moved on. Their lives, like the decaying logs, existed in frag-
ments—stone walls, cellar holes, and a regrown forest.[44]

Research at the experimental forest, Vincent Miller writes, "is breathtak-
ing." A professor of Catholic theology and culture at the University of Dayton,
he offered insights about "one of the most studied ecosystems in the world," a
place where theology can learn, looking with scientists at complex ecosystems.
Old-growth forests are disorderly and wild, huge trees centuries old where life
exists in many layers, a hint of timescales "beyond human reckoning." A single
tree "hosts hundreds of plants and animals and depends on countless ecological interconnections." Once referred to as "decadent forests," old growth are
places where life and death intermingle in a profusion of lichen, ferns, moss,

fungi, and dead snags, and centuries-old trees that seemingly reach out of sight.[45]

Miller traced the postwar story of converting old-growth forests to uniform, simplified, fast-growing young stands—clear-cutting, burning slash, and spraying herbicides to eliminate competing brush. Forest managers wanted timber, "but they also wanted order: a simple order they could understand and calculate." Andrews scientists, on the other hand, developed different perceptions, valuing old-growth forests for their complexity and as ecosystems with great diversity. Although the importance of lichen, moss, and decay was not obvious to casual observers, scientists discovered the hidden links and cycles of life in old forests, developing an understanding of larger ecosystems. To the theological scholar, scientists were involved in a "struggle to attend to the fullness of creation." Miller likened the old-growth forest to "a massive silent symphony of nutrient flows through thousands of plants, fungi, insects, bacteria, and animals."[46]

Ann Rosenthal's stay at the Andrews originated via a train trip between Portland and Seattle when she joined (unbeknown to her) Judith Li, an OSU fisheries scientist and author of *Ellie's Log*, at a table in the Café Car. She learned that Li was a semiretired professor in the university's Fisheries and Wildlife Department who was interested in collaboration between the arts and sciences. Rosenthal, an environmental artist, recalled that "a lively conversation ensued," which led to her learning about the Long-Term Ecological Reflections program. That chance encounter led to her residency at the experimental forest in late September 2018. The night before Rosenthal drove to the Andrews, she spent the dinner hour with Judy Li and her husband, delighted to enjoy conversations with scientists who thought in interdisciplinary terms. The next day, Fred Swanson treated her to a tour of research sites, including the surprising log-decomposition location. Thinking of it as "an operating theater," she found the setting attractive, "mysterious," and primeval, moss everywhere, as well as spongy footing underneath, a "carpet of living detritus." For the first two days, she worked on the Becoming Botanical project, a collaboration with a friend in India. She then turned her attention to "making work about this place."[47]

Rosenthal spent her remaining time sketching, painting, and taking photographs. She drew a branch with lichens attached, painting the colors a "dusty grey-green," and attempted to fuse the colors and textures of a decaying leaf in the dim lighting of her apartment. She continued working on the lichen paintings, piecing one of them together with a twig drawing "and really liked the combination." Although she failed to complete any projects during her stay, Rosenthal believed she had developed ideas for new work. She treasured the local trails, the water, and loved being "immersed in nature." On her last full day, Rosenthal compiled a list of "my take-aways": daily studio work (only or an hour or two); spend more time in nature, notice seasonal change, "and record them through art/photography"; draw more, especially out of doors; explore connections between realism and abstraction; get to know western PA; "Be happy, love life and nature, gratitude." On her way home, she enjoyed "wonderful, wet ride" to Salem, stopping at a waterfall, "the headwaters of the McKenzie River."[48]

Until Lissy Goralnik and three colleagues (two affiliated with the Andrews) published an article on arts and humanities projects at LTER sites in 2015, there had been no assessment of such efforts across the LTER Network. Goralnik's team surveyed the twenty-four LTER sites in 2015 and found that nineteen of them valued inquiries involving the arts and humanities. The Andrews Forest, Bonanza Creek (Alaska), North Temperate Lakes (Wisconsin), and several others formed Ecological Reflections in 2010, a loosely organized group of LTER sites whose members sponsored science/humanities programs. Goralnik and her coauthors conducted a survey of participating sites with guiding questions: What kinds of programs existed? How do sites value their science/humanities efforts? What challenges do they face? The responses pointed to limiting factors—funding, inadequate labor, and the availability of people with expertise to engage in such enterprises.[49]

Representing an expansive array of biomes "from conifer forests to grasslands, tundra to coral reefs," the LTER sites had produced a significant body of artistic and written work, some of which had been displayed at NSF headquarters in Washington, DC. The Goralnik review reported that six LTER sites consistently sponsored arts and humanities inquiries, including the "long-running writers-in-residence program at the Andrews Forest." Fifteen sites reported that they occasionally hosted programs, and three indicated they made no effort. The survey revealed that the most popular genres were painting (fifteen sites), photography (ten sites), and writing (eight sites). Nineteen of the twenty-four sites, however, solidly supported such efforts.[50]

The challenges and problems with ecological reflections programs were common across most sites. The most consistent shortcoming was funding, especially at locations where scientific investigations were already overtaxing staff. (It should be noted that the LTER Network office supported only a couple of workshops.) In a follow-up study published in 2017, Goralnik and colleagues found that the most active ecological reflections sites sponsored the following programs: writers' residencies; research experiences for under-graduate students; art and ecology workshops for teachers; performing and visual exhibits; field trips for visiting artists; historical research; and museum exhibits. Because there were great variations among the programs, collabora-tions across network sites also differed. In rationalizing the value of science/ humanities collaboration, Goralnik and colleagues cited their potential to "spark political action" in the face of the "advocacy limitations of science." At the time the authors published their review, little empirical work existed to validate "the value of arts-science collaboration or the impact of creative inquiry in ecological contexts."[51]

Frederick J. Swanson, who, with Kathy Moore, initiated the H. J. Andrews LTER Reflections program and influenced similar enterprises elsewhere, believes that science and humanities collaboration is directly related to navigating an uncertain future. The objective of the Long-Term Ecological Reflections program is to provide the public with information to make value judgements about natural resources. The Long-Term Ecological Research programs are similarly empowered to provide land managers with knowledge to make prudent management decisions. Both the LTER science program and the Reflections initiatives are exploratory. "We had no way of knowing," according to Swanson, "where the early work on northern spotted owls and old growth would take us in terms of science questions and social impact. Explore, be patient," he urged. Mixing arts, humanities, and science is part of a larger portfolio of inquiry involving academic institutions and governments. There is a need for "place-based, highly interdisciplinary teams taking the long view as those communities and places move through time in changing envi-ronmental (e.g., climate) and societal landscapes."[52]

In an address to the National Council on Science and the Environment in 2005, National Science Foundation president Arden Bement addressed the accomplishments of long-term ecological research and singled out for spe-cial praise the accomplishments of the LTER Reflections program. With the science-based LTER program celebrating its twenty-fifth anniversary, Bement made reference to an Andrews decomposition site and the essayist Robert Michael Pyle who found profound meaning in the place: "Most of us take the short-term view, most of the time," Bement quoted Pyle. Taking the long view required "faith in the future—even if you won't be there to see it for yourself."[53]

The Andrews Reflections program clearly profiles a powerful intersection between science and the arts and the humanities. The creative literary and art-work generated through the humanities and arts initiative provides important insights about place, time, beauty, awe-inspired wonder, and the investigations carried out on the mountainsides and the streamlets that feed Lookout Creek. In the decades following the creation of the experimental forest in 1948, scientists were the first to prowl the steep slopes of the watershed. The photographic records in the Andrews digital collections reveal a variety of activities—road-building, harvested units, and the complex research and observations that followed. Until the Reflections initiative and the work of creative writers, the public gleaned most of its information about northern spotted owls and similar phenomenon through journalistic coverage (interviews with scientists such as Jerry Franklin), news accounts, and magazine articles. That dialectic began to change when Alison Deming accompanied Ackers on a venture to the South Fork of the McKenzie to seek a spotted owl, an experience that Deming remembered as "beauty enhanced, not diminished by science."

The intersection between science and the humanities on the Andrews Experimental Forest also mirrors the long-ago ideas of Alexander von Humboldt (1769–1859) whose interests bridged the natural sciences with the social sciences and humanities, contending that the humanities deepened and broadened our understanding of the natural world. Humboldt's vision linked science with aesthetics—artistic productions, poetic creations, and literature—to extend and expand the meaning of human sensual relations with nature. The Andrews Experimental Forest website refers to the Long-Term Ecological Reflections program as an "Arts, Humanities, and Science Alliance," wherein interdisciplinary people from many fields gather "to reflect on the meaning and significance of the ancient forest ecosystem, as the forest—and its relation to human culture—evolves over time."[54]

Conclusion

> *Human Beings are not exempt from the iron law of species interdependency.*
> *We were not inserted as ready-made invasives into an Edenic world. Nor*
> *were we intended by providence to rule that world. The biosphere does not*
> *belong to us; we belong to it.*
>
> —Edward O. Wilson[1]

Enormous changes have taken place in the United States and globally since Roy Silen first piloted his way on foot through the huge forests in newly minted Blue River Experimental Forest in the summer of 1948. In his words, the Lookout Creek drainage was "untouched, a virgin area," a largely unmapped landscape of enormous four-hundred-year-old Douglas-fir and steep dissected slopes, with patches of younger trees along ridgetops.[2] Today, the H. J. Andrews Experimental Forest is roaded, approximately 25 percent of the forest was cut in the 1950s and 1960s, and the remainder of the Lookout Creek drainage has been left in mature-age-class trees. The headquarters site on the lower watershed now provides the amenities of a small community: dormitories, apartments, meeting halls, a library, and an outdoor covered pavilion. Yet, with the exception of scientists, the rare presence of a politician, and more recently an infusion of writers and artists, the Andrews remains a cloistered place, unknown to the vast majority of Oregon citizens.

Jerry Franklin has enjoyed long and affectionate ties to the Andrews Forest that date from the summer of 1957 when he worked on the Lookout Creek drainage as an undergraduate. Fred Swanson, who joined the Andrews group in 1972, views Franklin as the "cornerstone" of ecosystem research on the experimental forest. He was critical to the forest's participation in the International Biological Program and making the transition to the National Science Foundation's launch of the Long-Term Ecological Research program. Even when he moved to the University of Washington in 1986, Franklin remained close to activities on the Andrews. More than any forest/ecosystem scientist in the Pacific Northwest, he has been and remains at the cutting edge of proposals to maintain old-forest habitats and to provide modest supplies of timber to local communities. Between 2011 and 2013, Franklin and others began focusing on natural regeneration of forests after major disturbances— wildfires, windstorms, landslides, and insect infestations. In forester's lingo this meant understanding the significance of early-successional or early-seral ecosystems on disturbance sites and paying attention to sustaining biodiversity and the overall health of forest ecosystems, factors that should be important to forest managers.[3]

Although Franklin had largely ignored early-seral landscapes in the past, his many years of engagement with the post-blast Mount St. Helens environment convinced him that early-seral landscapes were critical to biodiversity. He began to appreciate the messy profusion of plant species after wildfires, shrub-like ecosystems that should be allowed to develop on their own. Willing to model a science that fit with political reality, Franklin created an early-seral management proposal that proposed logging two-thirds of the trees on a unit and leaving the area to recover naturally. Although scientists generally agreed that early-seral landscapes were important, Warren Cornwall cautioned in *Science* (2017) that there was considerable debate among forest managers about the efficacy of Franklin's scheme. When Franklin's proposal gained the attention of Oregon senator Ron Wyden, however, critics from all sides jumped in to the fray.[4]

Forest ecologist Dominick DellaSala, a coauthor of an article on early-seral forestry and head of the Geo Institute in Ashland, thought Franklin had gone too far, that logging was "a poor substitute for natural disturbances," which left a jumble of live and dead trees. Once a coauthor and now a strident critic, he thought Franklin's plan offered only "a veneer of environmental respectability" for logging on national forests. Ashland-based Andy Kerr charged that Franklin's proposal was "aimed at getting more wood for mills." When Franklin and colleague Norm Johnson collaborated with the Bureau of Land Management to experiment with a small logging unit near Roseburg, protesters occupied the site for nearly two years before a judge ruled against BLM for not fully examining potential environmental impacts.[5]

Amid the debates about using timber harvests to create early-seral landscapes—and the annual smokes billowing from summer wildfires in the Pacific Northwest—Jerry Franklin, Norm Johnson, and Debora Johnson were making the final revisions to their magisterial 646-page *Ecological Forest Management*, released in July 2018. Advertised as a textbook, its objective was to integrate decades of research in forest management to maintain healthy forest ecosystems and to meet the economic and social needs of humans. The global reach of *Ecological Forest Management* and its principles steeped in science on the Andrew, the authors believed, could be applied across most forest environments. The volume shuns the agricultural model of forest management, favoring natural ecological representations and offering a persuasive argument to encourage a reconsideration of harvest strategies on private as well as public lands. The text makes clear the many ways that forests benefit healthy habitats for diverse species, clean water, and carbon sequestration. The book counsels longer harvest rotations, leaving woody debris on the ground, and allowing some landscapes to grow naturally.[6]

The reviews of *Ecological Forest Management*, from widely different spectrums, have been favorable. V. Alaric Sample, president emeritus and senior fellow at the Pinchot Institute for Conservation and adjunct professor of environmental science and policy at George Mason University, reviewed the book for the *Journal of Forestry*, citing it as "one of the landmarks of forestry literature," a chronicle that begins and ends with "statements about what ecological forest management is *not.*" The lengthy centerpiece to the book, according to Sample, includes "well-documented chapters exploring forests as ecosystems"; addressing issues of forest dynamics, succession, natural and human disturbances; and illustrating harvest strategies that leave "green trees, snags, and logs in late-successional forests." Andy Kerr, a self-described "card-carrying conservationist" and environmental radical, praised Franklin and Johnson for firing a shot across the bow of conventional plantation forestry, exposing its limits and reflecting radical changes in forestry, "a profession that is part science, part art, and part craft." The book thoroughly debunks the agricultural model, once a cherished principle in colleges of forestry. Kerr concluded, "The forests of the world will be better off for the publication and utilization of *Ecological Forest Management*."[7]

Mount St. Helens continues to influence the thinking of Andrews scientists. The Forest Service began sponsoring a group of scientists who met every five years for weeklong campouts to check on permanent reference plots and gather information on the varying response rates of plant and animal life. The scientists published an assessment on the twenty-fifth anniversary of the eruption and another summary commemorating thirty-five years of post-event investigations in 2017. Fred Swanson joined Charles Crisafulli, a research ecologist with the Pacific Northwest Research Station, in writing the concluding chapter to *Ecological Responses at Mount St. Helens: Revisited 35*

Years after the 1980 Eruption, reviewing the contributions of Mount St. Helens research. The authors contend that the Mount St. Helens eruption paralleled the emergence of disturbance ecology and has advanced our understanding of volcano ecology. With Crisafulli in charge, the Mount St. Helens ecological observatory is renowned for its dedicated scientific and monitoring program and public outreach.[8]

Twenty-first-century research on the experimental forest has accelerated ecological management on national forests in the Pacific Northwest, underscoring again the environmental contributions of the Andrews Forest. Reducing forest fragmentation, honoring natural fire regimes, removing nonnative species, and improving habitats for threatened and endangered aquatic and terrestrial species has taken precedence over timber extraction. Despite political opposition in the halls of Congress and with the current administration, respecting biodiversity has gained traction. Yet, for all the efforts to slow the loss of biodiversity and provide strategies to confront the threat of a warming climate, "the world is in desperate condition," according to Edward O. Wilson, author of the Pulitzer Prize–winning book, *Half-Earth: Our Planet's Fight for Life.* Because the Earth is experiencing sharp declines in diversity, Wilson believes there is a need for "a major shift in moral reasoning."[9] Although the Andrews Forest represents only a microcosm on the Earth's surface, its research has influenced management decisions on the national forest system in the Pacific Northwest and beyond. What, then, is the significance of the Andrews in our present day?

Scientists who have been affiliated with the Andrews Forest for many years offer differing perspectives about its larger meaning. To Mark Harmon, who pioneered the log-decomposition studies, the experimental forest is "a place people care about," an aesthetic involving a community of scientists who joined together to participate in significant research. Even at Edward O. Wilson's global scale, Harmon believes, "you need to care about place." One of the strengths of the Andrews participation in the LTER is its record-keeping, mechanisms for scientists to find information. Harmon stresses the importance of the diverse community of scientists on the Andrews and values the current social science research on the forest and the visiting writers and artists.[10] Recently retired as a forest ecologist from the Pacific Northwest Research Station in Corvallis, Tom Spies agrees that the Andrews is a mere speck on a planetary scale. Despite its small size, however, the forest has made significant contributions to ecosystems science at regional, national, and global scales. It is "a small place with big ideas." Spies applauds the Andrews for pitching the

Mark Schulze, forest director of the H. J. Andrews Experimental Forest since 2008, lives at the headquarters on lower Lookout Creek and oversees field stations, research infrastructure, educational activities, and visitations.

relevance of the "real world" in its reports to the National Science Foundation, and, like Mark Harmon, he considers the sense of community an important component of its research agenda.[11]

Barbara Bond, the only woman to serve as principal investigator for the experimental forest's Long-Term Ecological Research (LTER) program (2008–2014), emphasizes the importance of global issues. The pressing problems of the world, she believes, need to be studied on a global scale. Describing herself as "not a place bound researcher," Bond believes that western Oregon provides a far wider and more appropriate field for effective ecological investigations. With the emergence of the Reflections program, she sees the forest as a place where artists and writers can think creatively about ongoing research.[12] Forest Director Mark Schulze indicates that the Andrews has always been a vehicle for scientists to investigate how humans should use landscapes sustainably to maintain functioning ecosystems. Its investigations have significantly influenced management on the Willamette National Forest. Schulze, who is the main point of contact for visitors to the Andrews, continues the forest's working relationships with school districts and science teachers. Because of its international reputation for ecosystem research, the forest also attracts visitors from abroad as well as forest managers wanting to learn more about the Andrews collaboration with the Willamette National Forest.[13]

Fred Swanson, whose publications and insights are sprinkled throughout this book, considers the Andrews Experimental Forest "a seedbed for discovery," an effort to see how the world works, studying land-use practices and

Michael Paul Nelson,
with graduate degrees in
philosophy, became the
Andrews principal investigator
for the National Science
Foundation's Long-Term
Ecological Research grants in
2012, the first non-scientist to
hold the position.

water/forest interactions. Swanson argues that the experimental forest has been a flagship site for scientific research involving ecosystem investigations and much more. It is critically important, he argues, to remember that land managers are the interface with society, because they "live in the real world."[14]

When Michael Nelson became lead principal investigator for the H. J. Andrews Experimental Forest in 2012, he was the first non-scientist appointed to the position. With a master's degree from Michigan State and a PhD from Lancaster University in England (both in philosophy), Nelson's mantra is to ask questions about the social significance of scientific research. During his tenure with the Andrews, he has come to see the forest as "a place of inquiry and research," a site for investigating the multifaceted consequences of climate change. He regards the Andrews as a mature research site, a place where scientists need to pursue surprise and move beyond old models.[15]

As this book goes to press (2020), Andrews aficionados can take pride in the forest's significant place in a Pulitzer Prize–winning book. Celebrated fiction writer Richard Powers, with a National Book Award in 2006, won the Pulitzer Prize for fiction in 2019 for *The Overstory: A Novel*, creative reflections on humans and the natural world. Although the publisher advertises the book as fiction, that argument falls flat in the presence of the fictional Dr. Patricia Westerford, who has a grant to join research teams from Corvallis at the Franklin Experimental Forest in the Oregon Cascades. The forest's headquarters site is described in the 1990s as crude, "a mildewed trailer in the

Ghetto in the Meadow. . . . The latrines and the common showers are sinful indulgences."[16]

Kathleen Dean Moore, cofounder of the Spring Creek and OSU Emeritus Distinguished Professor of Philosophy, reviewed *Overstory* for the Corvallis–Benton County Library's Random Review series in January 2019. The book, in her view, might be called "historical fiction about our own place and our own histories." We know the characters and have hiked the trails in this book, "we have smelled the setting, breathed its air, tried to wrap our arms around it—the damp forest and shafting light." The book is brilliant, the characters are alive, complex, and have lived in larger contextual worlds. The novel branches out like a tree, Moore argues, a story told "through one person's eyes, then the next." The Franklin Forest is about science, the "newest science about trees." *Overstory* is important, Moore believes, because we know more about forests now than we did thirty or forty years ago, and we now prize old-growth forests. The book "invites us to imagine how people [like us] find courage to stand up for what they love too much to lose."[17]

The research accomplishments on the Andrews Forest for the last half century are relevant to a recent United Nations publication released in May 2019, the *Intergovernmental Science-Policy Platform on Biodiversity and Ecosystem Services for the Americas* (IPBES). Although the report revealed widespread evidence of stressed and degraded global ecosystems and loss of biodiversity, it concluded that the Western Hemisphere was better equipped to contribute to sustainable life for humans and other species than most other places. The story of the Americas, however, also indicated that most countries were "using nature at a rate that exceeds nature's ability to renew the contributions it makes to the quality of life." Wasteful use of surface and subsurface water was diminishing supplies, and biodiversity and ecological systems were deteriorating. The causes of those downward-trending conditions are population increases, uninhibited economic development, and "weak governance systems and inequity." Climate change, the IPBES reported, was amplifying those forces. To slow or reverse the degradation of ecosystems, the assessment recommended increasing protected areas, restoring sullied ecosystems, upgrading the sustainability of lands outside the bounds of reserved set-asides, and "mainstreaming conservation and sustainable use of biodiversity in productive sectors."[18]

To better appreciate the dire threats to biodiversity and human livelihoods, the IPBES called for a better understanding of the historical underpinnings and drivers of environmental change. Although the H. J. Andrews

Experimental Forest is but a microcosm of the Earth's surface, its portfolio of research achievements positions it to provide a model for understanding healthy forest/stream relations, ecosystem restoration, and the conditions that produce clear water for consumers. Andrews research provides strategies to sustain quality habitats for species, maintain biodiversity, and manage stands of trees for carbon sequestration and water supplies. If the agricultural model of forestry is faulty, experiments on the Lookout Creek drainage reveal the shortcomings of clear-cuts, suggesting selective harvesting to replicate a semblance of biodiversity and to achieve desirable water yields. To illustrate the common features between the IPBES recommendations for the Americas and investigations on the Andrews Forest, the latter offers on its 15,800 acres both protected areas and sectors devoted to ecological restoration and experimentation. In addition, Andrews personnel have enjoyed many years of collaborative relations with the Willamette National Forest and other federal forests in the region, offering expertise in maintaining viable aquatic systems and ecologically healthy approaches to harvesting trees.

These are the challenges ahead for the Andrews Experimental Forest. During the summer of 2017, the nearby Horse Creek Complex of fires posed a potential threat to the Andrews headquarters, and staff and visitors were put on notice that they might have to evacuate. The following year, amid the forest celebrating its seventieth anniversary, Oregon witnessed yet another alarming wildfire season, including the Terwilliger Fire on the South Fork of the McKenzie River. Those wildfires, and others across Oregon, took place in the midst of longer, hotter, drier summers, a trend expected to accelerate—even in the moist, temperate climate of the western Cascades. The expectation is that warmer summers will be favorable for the spread of pests and pathogens, in addition to the potential for even more wildfires. For a small corner of the globe, however, the Andrews Forest possesses a superb infrastructure of seven decades of historical record-keeping: data on temperatures, seasons, precipitation patterns, species adjustments, major disturbances, and their association with human activities. The forest's rich database and its well-documented records provide a wealth of historical information for assessing what many believe will be a very different future.

Appendix A

Long-Term Ecological Research (LTER) Grants
and Principal Investigators for Each

FIVE-YEAR CYCLES

LTER1	1980–1985	Richard Waring
LTER2	1986–1990	Jerry Franklin
LTER3	1991–1996	Fred Swanson

SIX-YEAR CYCLES

LTER4	1996–2002	Fred Swanson
LTER5	2002–2008	Mark Harmon
LTER6	2008–2014	Barbara Bond
LTER7	2014–2020	Michael Nelson

Notes

INTRODUCTION

1 Quotation on the cover of *Silent Spring Review: Researching the Environment and Women's Health* (Fall 2000).

2 For purposes of brevity, I use its present name, the Pacific Northwest Research Station, in lieu of the agency's earlier title, the Pacific Northwest Forest and Range Experiment Station.

3 William G. Robbins, *Lumberjacks and Legislators: Political Economy of the U.S. Lumber Industry, 1890–1941* (College Station: Texas A&M University Press, 1982), 19–20, 26–27; Robbins, *American Forestry: A History of National, State, and Private Cooperation* (Lincoln: University of Nebraska Press, 1985), 5–6, 7, 9; and Robbins, "Willamette National Forest," *Oregon Encyclopedia*, https://oregonencyclopedia.org/.

4 Robbins, *Lumberjacks and Legislators*, 30–32; and Robbins, *American Forestry*, 11–16.

5 Robbins, *American Forestry*, 13–19; and Margaret Herring and Sarah Greene, *Forest of Time: A Century of Science at Wind River Experimental Forest* (Corvallis: Oregon State University Press, 2007), 18–22.

6 Herring and Greene, *Forest in Time*, 36–38, 155–156; and Ivan Doig, *Early Forest Research: The Story of the Pacific Northwest Forest and Range Experiment Station* (Washington, DC: USDA Forest Service, 1976), 1, 4.

7 Thornton T. Munger, Report of the Pacific Northwest Forest Experiment Station for the Calendar Year 1936, USDA Forest Service, R-NW, Reports, Annual, January 17, 1937; Pringle Falls Experimental Forest and Research Natural Area, https://www.fs.fed.us/pnw/exforests/pringle-falls/; Cascade Head Experimental Forest, https://www.fs.fed.us/pnw/exforests/cascade-head/; and Starkey Experimental Forest and Range, https://www.fs.fed.us/pnw/exforests/starkey/, all in USDA Forest Service, Pacific Northwest Research Station. For the discontinued Port Orford Cedar Experimental Forest, see https://www.na.fs.fed.us/spfo/pubs/silvics_manual/Volume_1/chamaecyparis/lawsoniana.htm.

8 William G. Robbins, *The People's School: The History of Oregon State University* (Corvallis: Oregon State University Press, 2017), 226.

9 See Robbins, "The H. J. Andrews Experimental Forest: Seventy Years of Pathbreaking Forest Research," *Oregon Historical Quarterly* 119, no. 4 (Winter 2018): 454–485.

10 James T. Callahan, "Long-Term Ecological Research," *BioScience* 34, no. 6 (June 1984): 363–364.

11 Sara Zaske, "Seeing the Forest through the Trees: H. J. Andrews Experimental Forest Tackles Tough Ecological Questions," *Focus on Forestry* 19, no. 2 (Spring 2006): 26–27; http://www.nasonline.org/about-nas/history/archives/collections/ibp-1964-1974-1.html; Frank B. Golley, *A History of the Ecosystem Concept in Ecology: More Than the Sum of Its Parts* (New Haven, CT: Yale University Press, 1993), 4, 110, 135–138; David C. Coleman, *Big Ecology: The Emergence of Ecosystem Science* (Berkeley: University of California Press, 2010), 24, 45, 48, 74; National Research Council, *Finding the Forest in the Trees: The Challenge of Combining Environmental Data* (Washington, DC: National Academy Press, 1995), 46–47.

12 Jon R. Luoma, *The Hidden Forest: The Biography of an Ecosystem* (New York: Henry Holt, 1999), 8.

13 Paul W. Hirt, *A Conspiracy of Optimism: Management of the National Forests since World War Two* (Lincoln: University of Nebraska Press, 1994), xxi–xxxii.

14 William G. Robbins, *Landscapes of Conflict: The Oregon Story, 1940–2000* (Seattle: University of Washington Press, 2004), 175–177.

15 Sally Duncan, "Openings in the Forest: The Andrews Story," *Forest History Today* (Fall 1999): 20–21; Ivan Doig, *Early Forest Research: The Story of the Pacific Northwest Forest and Range Experiment Station* (Washington, DC: USDA Forest Service, 1976), 19, 21; and Carl M. Bernstein and Jack Rothacher, *A Guide to the H. J. Andrews Experimental Forest* (USDA Forest Service, Pacific Northwest Forest and Range Experiment Station, 1959), 1–5, 7–17.

16 Duncan, "Openings in the Forest," 20–21; and Doig, *Early Forest Research*, 19, 21.

17 "Forest Report Asks Research," *Oregonian*, July 13, 1951; and Jerry Dunford, Work Plan for Forest Influences Studies: H. J. Andrews Experimental Forest, June 13, 1953. Both items found in William G. Robbins Papers, Special Collections and Archives, Oregon State University (hereafter SCARC).

18 Zaske, "Seeing the Forest through the Trees," 26–27; National Academy of Sciences, Organized Collections, "The International Biological Program (IBP), 1964–1974," http://www.nasonline. org/about-nas/history/archives/collections/ibp-1964-1974-1.html; Golley, *A History of the Ecosystem Concept*, 4, 110, 135–138; Coleman, *Big Ecology*, 24, 45, 48, 74; and National Research Council, *Finding the Forest in the Trees*, 46–47.

19 Golley, *A History of the Ecosystem Concept*, 6; Coleman, *Big Ecology*, 101–103, 121.

20 See the following Andrews documents for comparing the shift in research: R-NW Branch Stations, Blue River Experimental Forest, Agreement, June 4, 1948; H. J. Andrews Experimental Ecological Reserve Proposal (1976); and Report of the National Advisory Committee of the H. J. Andrews EER Site Visit (September 1978), all in SCARC.

21 For the extended discussion involving the spotted owl and the Dwyer decision, see William G. Robbins, *Landscapes of Conflict: The Oregon Story, 1940–2000* (Seattle: University of Washington Press, 2004), 206–209, 321.

22 Daniel Nelson, *Nature's Burdens: Conservation and American Politics, the Reagan Era to the Present* (Logan: Utah State University Press, 2017), 7, 28–29; George Hoberg, "Science, Politics, and U.S. Forest Service Law: The Battle over the Forest Service Planning Rule," *Natural Resources Journal* 44 (Winter 2004): 2–8; and Hirt, *A Conspiracy of Optimism*, xxxvii–xxxviii.

23 Hoberg, "Science, Politics, and U.S. Forest Service Law," 4–18; and Nelson, *Nature's Burdens*, 28–29.

24 Nelson, *Nature's Burdens*, 45–48.

25 Michael E. Kraft and Norman J. Vig, "Environmental Policy in the Reagan Presidency," *Political Science Quarterly* 99, no. 3 (Autumn 1984): 433–435, 437–438; Erik D. Olsen, "The Quiet Shift of Power: Office of Management and Budget Supervision of Environmental Protection Agency Rulemaking under Executive Order 12,291," *Virginia Journal of Natural Resources Law* 4, no. 1 (1984–1985): 3–6, 80; and Thomas O. McGarity, "Regulatory Reform in the Reagan Era," *Maryland Law Review* 45, no. 2 (1986): 270–272; http://heinonline.org/HOL/LandingPage?handle= hein.journals/mllr45&div=17&id=&page=.

26 Anna Maria Gillis, "The New Forestry: An Ecosystem Approach to Land Management," *BioScience* 40, no. 8 (September 1990): 558562; *Forest Ecosystem Management: An Ecological, Economic, and Social Assessment; Report of the Forest Ecosystem Management Assessment Team* (USDA and US Department of the Interior, July 1993); and Jack Ward Thomas et al., "The Northwest Forest Plan: Origins, Components, Implementation Experience, and Suggestions for Change," *Conservation Biology* 20, no. 2: 277–287.

27 R. Edward Grumbine, *Ghost Bears: Exploring the Biodiversity Crisis* (Washington, DC: Island Press, 1992); and Alston Chase, *In a Dark Wood: The Fights over Forests and the Rising Tyranny of Ecology* (Boston: Houghton Mifflin, 1995).

28 Gene E. Likens, "Limitations to Intellectual Progress in Ecosystem Science," in *Successes, Limitations and Frontiers in Ecosystem Science*, edited by P. M. Groffman and M. L. Place (New York: Springer, 1998), 241, 248–249.

29 David R. Foster, ed., *Hemlock: A Forest Giant on the Edge* (New Haven, CT: Yale University Press, 2014), xxv.

CHAPTER 1

1 H. J. Andrews and R. W. Cowlin, *Forest Resources of the Douglas-Fir Region* (Washington, DC: GPO, 1940), 1.

2 *Oregonian*, March 26, 1951.

3 *Oregonian*, March 26, 1951; H. J. Andrews Resume, circa 1938; Earle Clapp to Andrews, November 6, 1929; and Thornton Munger to Andrews, November 20, 1929, all in SCARC. For a brief

history of the forest survey, see Ivan Doig, "The Beginning of the Forest Survey," *Journal of Forest History* (January 1976): 21–27.

4 Clapp to Munger, March 4, 1935, in SCARC. For Earle Clapp's résumé, see "Earle H. Clapp (1877–1943): Acting Chief of the Forest Service, 1939–1943," US Forest Service Headquarters Collection, Forest History Society, https://foresthistory.org/research-explore/us-forest-service-history/people/chiefs/earle-h-clapp-1877-1970/. In addition to pushing progressive policies during his long tenure with the agency, Clapp also authored one of the most incisive documents ever produced by the Forest Service, the Copeland Report, *A National Plan for Forestry* (1933), named after its principal US Senate sponsor, Senator Royal Copeland of New York. For a description of the Copeland Report, see William G. Robbins, *Lumberjacks and Legislators: Political Economy of the U.S. Lumber Industry, 1890–1941* (College Station: Texas A&M University Press, 1982), 168–170, 177–178.

5 H. J. Andrews, "Land Classification as a Solution to Wasteful Land Uses," presented to the Oregon Planning Council, October 27, 1934, in SCARC. For an account of Roosevelt's Natural Resources Planning Board, see Marion Clawson, *New Deal Planning: The Natural Resources Planning Board* (Washington, DC: Resources for the Future, 1981).

6 Andrews, "Integration of Land Uses," presented to the Second Pacific Northwest Planning Conference, December 10, 1934, in SCARC.

7 Samuel Trask Dana to Andrews, July 6, 1938; Andrews to Dana, December 8, 1938; H. J. Andrews (photocopy) "Research Necessary to Forestry's Contribution to National Welfare," *Michigan Forester* (1939): 19–21, 66–67; and Lyle Watts to Andrews, February 22, 1943, all in SCARC.

8 Andrews to Dana, July 17, 1939; "Andrews Joins Regional Forester Staff, North Pacific Region," *Forest News*, nd; and Lyle Watts to Andrews, February 22, 1943, all in SCARC. Clapp succeeded Chief Forester Ferdinand Silcox, who died suddenly in 1939. See Robbins, *Lumberjacks and Legislators*, 220. The other two missions of the Forest Service are national forests and research.

9 Robbins, *Lumberjacks and Legislators*, 168–170, 177–180, 220, 234; and Andrews to Clapp, January 24, 1945, in SCARC.

10 Max Geier, *Necessary Work: Discovering Old Forests, New Outlooks, and Community on the H. J. Andrews Experimental Forest, 1948–2000*, USDA Forest Service, Pacific Northwest Research Station, General Technical Report, PNW-GTR 687 (March 2007), 9–18. National forest timber production increased from 4.4 billion board feet in 1952 to more than 12 billion board feet in 1966. See Burnett and Davis, "Getting Out the Cut: Politics and National Forest Timber Harvests, 1960–1995," *Administration and Society* (2002): 202–206, https://doi.org/10.1177/0095399702034002004.

11 R-NW, Branch Stations, Blue River Experimental Forest, Agreement, June 4, 1948, in SCARC; and Ivan Doig, *Early Forest Research: The Story of the Pacific Northwest Forest and Range Experiment Station* (Washington, DC: USDA Forest Service, 1976), 19. The term "working circle" refers to a Forest Service plan in which national forest timber sales would be designated for local mills. Adopted at the end of the Second World War, the working circle strategy was abolished by 1970.

12 R-NW, Branch Stations, Blue River Experimental Forest, Agreement, June 4, 1948.

13 Jon R. Luoma, *The Hidden Forest: The Biography of an Ecosystem* (New York: Henry Holt, 1999), 11; and Blue River Experimental Forest, Representing the Old-Growth Douglas-Fir Type of the Central and Southern Cascades, Willamette National Forest, Oregon, R-NW, Region 6, US Forest Service, approved by Acting Chief R. E. McCardle (July 28, 1948), 1, 7, in SCARC.

14 Sally Duncan, "Openings in the Forest: The Andrews Story," *Forest History Today* (Fall 1999): 20–21; Doig, *Early Forest Research*, 19, 21; and Establishment Report, Blue River Experimental Forest, USDA Forest Service, R-NW, Branch Stations, Blue River Experimental Forest, all in SCARC.

15 Blue River Experimental Forest, Representing the Old-Growth Douglas-Fir Type, 7; and Working Plan for Blue River Harvest Cuttings—Sale #1, USDA Forest Service, RS-NW, Blue River Sale, 1–2, all in SCARC.

16 Working Plan for Blue River Harvest Cuttings—Sale #1, USDA Forest Service, RS-NW, Blue River Sale, 15–16, in SCARC.

17 Max Geier, interview with Roy Silen, September 9, 1996; and Robert H. Ruth and Roy R. Silen, *Suggestions for Getting More Forestry in the Logging Plan*, PNW Old Series Research Notes No. 72, Pacific Northwest Forest and Range Experiment Station (Portland: USDA Forest Service, 1950), 1–19.

18 Geier, interview with Silen.

19 Roy Silen, "Facts about the H. J. Andrews Experimental Forest," USDA Forest Service, H. J. Andrews Experimental Forest (August 1953), 1–6; and R-NW, Branch Stations, H.J. Andrews Experimental Forest, *Memorandum of Understanding*, April 30, 1953, in SCARC.

20 *The Timberman* (May 1951): np; *Journal of Forestry* (May 1951); and William O. Douglas to Mrs. Andrews, April 3, 1951, all in SCARC.

21 Program, Dedication of H. J. Andrews Experimental Forest, July 26, 1953; and *Oregon Journal*, July 23, 1953, both in SCARC.

22 *Oregonian*, July 13, 1951; and Jerry Dunford, Work Plan for Forest Influences Studies: H. J. Andrews Experimental Forest, USDA Forest Service, June 13, 1953, in SCARC.

23 Max Geier interview with Roy Silen, September 9, 1996, in SCARC.

24 Memorandum of Understanding for Cooperative Use of Facilities at the Blue River Ranger Station, April 17, 1957, R-NW, Branch Stations, Willamette Research Center, in SCARC.

25 R-NW, Branch Stations, H. J. Andrews Experimental Forest, Memorandum of Understanding, January 3, 1959, in SCARC.

26 Jack S. Rothacher, "A Project Plan to Study the Effects of Road Building and Logging on Erosion Sedimentation and Streamflow Behavior on the H.J. Andrews Experimental Forest," USDA Forest Service, RI-NW, Research Program, Project Plans, May 23, 1958, in SCARC.

27 Carl M. Bernstein and Jack Rothacher, *A Guide to the H. J. Andrews Experimental Forest*, USDA Forest Service, Pacific Northwest Forest and Range Experiment Station (1959), 1–5, 7–17, in SCARC.

28 Jack Rothacher, "Working Plan for Small Watershed Study," USDA Forest Service, RI-NW, Watershed Management, Small Watersheds, Water Quality and Yield (February 4, 1958), 1–5, in SCARC.

29 Jack Rothacher, "Working Plan for Small Watershed Study," USDA Forest Service, RI-NW, Watershed Management, Small Watersheds, Water Quality and Yield (February 4, 1958), 6–8, in SCARC.

30 Rothacher, "Working Plan," 8, 12–16.

31 "Logging and Forestry Know-How," *Forest Industries* 89, no. 2 (1962): 62–63.

32 Geier interview with Ted Dyrness, September 11, 1996; and Geier interview with Jerry Franklin, September 13, 1996, in SCARC.

33 Geier interview with Al Levno, September 12, 1996.

34 Memorandum of Understanding, R-NW, Branch Stations, H. J. Andrews Experimental Forest, January 3, 1959, in SCARC.

35 E. G. Dunford to Jack Rothacher (Speed-Memo), March 14, 1962; Rothacher reply, March 16, 1962; and Dunford to Rothacher, United States Government Memorandum, April 3, 1962, in SCARC.

36 Max Geier, interview with Ted Dyrness, September 11, 1996; and Geier, interview with Jerry Franklin, September 13, 1996.

37 Geier interview with Dyrness, September 11, 1996; Geier interview with Andrews Group, September 22, 1997; Geier, *Necessary Work*, 109–110; and R. L. Fredriksen, "Christmas Storm Damage on the H. J. Andrews Experimental Forest," *US Forest Service Research Note*, Pacific Northwest Forest and Range Experiment Station (Portland, August 1965), in SCARC.

38 Geier interview with Levno, September 12, 1996; and Geier, *Necessary Work*, 104–109.

39 Geier interview with Dyrness, September 11, 1996; and C. T. Dyrness, *Mass Soil Movements in the H. J. Andrews Experimental Forest*, USDA Forest Service, Pacific Northwest Forest and Range Research Station, Research Note, PNW-42 (Portland, 1967), 1–2, 4–5, 12.

40 R. L Fredriksen, "Sedimentation after Logging Road Construction in a Small Western Oregon Watershed," *Proceedings of the Federal Inter-Agency Sedimentation Conference*, USDA, Miscellaneous Publications 970 (1965): 56–59.

41 Jerry F. Franklin, Frederick C. Hall, C. T. Dyrness, and Chris Maser, *Federal Research Natural Areas in Oregon and Washington: A Guidebook for Scientists and Educators*, Pacific Northwest Forest and Range Experiment Station, Portland; Stephanie A. Snyder, Lucy E. Tyrrell, and Robert G. Haight, "An Optimization Approach to Selecting Research Natural Areas in National Forests," *Forest Science* 45, no. 3 (August 1999): 458–459; and Rezneat M. Darnell, "Natural Area Preservation: The US/IBP Conservation of Ecosystems Program," *BioScience* 26, no. 2 (February 1976): 105–108.

42 Sarah E. Greene, Tawny Blinn, and Jerry F. Franklin, *Research Natural Areas in Oregon and Washington: Past and Current Research and Related Literature*, USDA Forest Service, Pacific Northwest Research Station, General Technical Report, PNW-197 (November 1986), 2–3.

43 Kevin R. Marsh, "'This Is Just the First Round': Designating Wilderness in the Central Oregon Cascades, 1950–1964," *Oregon Historical Quarterly* 103, no. 2 (Summer 2002): 210–215.

44 Marsh, "This Is Just the First Round," 216–229.

45 Kevin R. Marsh, *Drawing Lines in the Forest: Creating Wilderness Areas in the Pacific Northwest* (Seattle: University of Washington Press, 2007), 99–103.

46 Marsh, *Drawing Lines*, 103–118. Noted environmental historian Samuel Hays believed that the French Pete Valley was the first completely forested drainage to be designated as wilderness and "to have a biological content that was taken seriously." See Samuel P. Hays, "The Trouble with Bill Cronon's Wilderness," *Environmental History* 1, no. 1 (1996): 29–32.

47 Gordon Grant, "Downstream Effects of Timber Harvest Activities on the Channel and Valley Floor Morphology of Western Cascade Streams," PhD diss., Johns Hopkins University, 1986. For publications that grew out of his dissertation, see Gordon Grant, "Sediment Movement at the Oregon LTER Site (H. J. Andrews Site)," in J. Rodgers, comp., *Sediment Movement at LTER Sites: Mechanics, Measurement, and Integration with Hydrology* (Champaign: State Water Survey Division, Water Section, University of Illinois), 4–9; Grant, "Assessing Effects of Peak Flow Increases on Stream Channels: A Rational Approach," in Robert Callaham and Johannes DeVries, coordinators, *Proceedings of the California Watershed Management Conference, 1986* (Berkeley: Wildland Resources Center, Division of Agriculture and Natural Resources, University of California, 1987), 142–149; and Grant, *The Rapid Technique: A New Method for Evaluating Downstream Effects of Forest Practices on Riparian Zones*, General Technical Report, PNW-GTR-220 (Portland: USDA Forest Service, Pacific Northwest Forest and Range Experiment Station, 1988); and J. F. Franklin and R. H. Waring, "Information on the H. J. Andrews Experimental Forest, Western Oregon Cascade Range," in *Coupling of Ecological Studies with Remote Sensing: Potentials at Four Biosphere Reserves in the United States*, edited by M. I. Dyer and D. A. Crossley (Washington, DC: Department of State, Bureau of Oceans and International Environmental and Scientific Affairs, 1986), 34–41.

48 Elena Aronova, Karen S. Baker, and Naomi Oreskes, "Big Science and Big Data in Biology: From the International Geophysical Year through the International Biological Program to the Long Term Ecological Research (LTER) Network, 1957—Present," *Historical Studies in Natural Science* 40, no. 2 (Spring 2010): 192, 196; and N. W. Pirie, "Introduction: The Purpose and Function of the International Biological Programme," *Proceedings of the Nutrition Society* 26, no. 1 (March 1967): 125–126.

49 Pirie, "Introduction," 125–126; Frederick E. Smith, "The International Biological Program and the Science of Ecology," *Proceedings of the National Academy of Sciences of America* 60, no. 1 (May 1968): 5–11; E. M. Nicholson, *Handbook to the Conservation Section of the International Biological Programme* (Oxford: Blackwell Scientific, 1968), 7, 15–16; and Aronova et al., "Big Science and Big Data in Biology," 199–200.

50 Coleman, *Big Ecology*, 16–17, 74; and Frank B. Golley, *A History of the Ecosystem Concept in Ecology: More Than the Sum of Its Parts* (New Haven, CT: Yale University Press, 1993), 1–2.

51 The Frederick Smith and W. F. Blair remarks are in Robert P. McIntosh, *The Background of Ecology: Concept and Theory* (New York: Cambridge University Press, 1985), 214–218.

52 McIntosh, *Background of Ecology*, 218–219.

53 Aronova et al., "Big Science and Big Data in Biology," 208–209.

54 H. J. Andrews Experimental Forest, undated mimeograph copy, in SCARC.

55 H. J. Andrews Experimental Forest—A Proposed Comprehensive Research Site, undated mimeograph copy; and Jack Rothacher to Frederick E. Smith, July 21, 1967, both in SCARC. Rothacher likely had access to a prepublication copy of the now classic Jerry Franklin and C. T. Dyrness, *Vegetation of Oregon and Washington*, first published in 1969. For the revised and expanded version see, Franklin and Dyrness, *Natural Vegetation of Oregon and Washington*, Pacific Northwest Forest and Range Experiment Station, USDA Forest Service, General Technical Report, Portland, 1973. For one of Jay Gashwiler's many small-mammal studies, see "Small Mammal Study in West-Central Oregon," *Journal of Mammalogy* 40, no. 1 (February 1959): 123–129.

56 Arthur McKee, "Focus of Field Stations," *Bulletin of the Ecological Society of America* 79, no. 4 (Oct. 1998): 241–242.

57 Jerry F. Franklin, H. J. Andrews Experimental Forest, Nomination of an IBP Study Area, Outline of Criteria, in SCARC; Franklin and Dyrness, *Vegetation of Oregon and Washington* (1969).

58 Geier, interview with Jerry Franklin, September 13, 1996; and Geier interview with H. J. Andrews International Biome Program group, February 10, 1998. An Illinois native, Richard War-

ing earned bachelor's and master's degrees from the University of Minnesota and his PhD at the University of California, Berkeley.

59 Geier, interview with H. J. Andrews International Biome Program group, 2–5; and Geier, interview with Franklin (September 13, 1996), 12.

60 E. M. Nicholson, *Handbook to the Conservation Section of the International Biological Programme* (Oxford: Blackwell Scientific, 1968), 17–18, 36; and McIntosh, *Background of Ecology*, 219.

61 For a summary of these findings, see S. V. Gregory, F. J. Swanson, W. A. McKee, and K. W. Cummins, "An Ecosystem Perspective on Riparian Zones," *BioScience* 41, no. 8 (1991): 540–551; and Victoria Ruiz-Villanueva and Markus Stoffel, "Frederick J. Swanson's 1976–1979 Papers on the Effects of Instream Wood on Fluvial Processes and Instream Wood Management," *BioScience* 48, no. 9: 681–689.

62 Geier interview with Mike Kerrick, August 28, 1996, in SCARC.

63 Jerry F. Franklin, L. J. Dempster, and Richard H. Waring, eds., *Research on Coniferous Forest Ecosystems: First Year Progress in the Coniferous Forest Biome, US/IBP*, Proceedings of a Symposium held at Northwest Scientific Association, Forty-Fifth Annual Meeting, Bellingham, Washington, March 23–24, 1972 (Portland: Pacific Northwest Forest and Range Experiment Station, 1972), Foreword and Contents.

64 Franklin, "Why a Coniferous Forest Biome?" 3–4.

65 Franklin, "Why a Coniferous Forest Biome?" 4–5.

66 Stanley P. Gessel, "Organization and Research Program of the Western Coniferous Forest Biome," in *Research on Coniferous Forest Ecosystems*, 7–14.

67 Franklin and Dyrness, *Vegetation of Oregon and Washington* (1969); and Franklin and Dyrness, *Natural Vegetation of Oregon and Washington* (1973). Research Natural Areas are representative of common ecosystems in their natural condition and serve as model baselines for reference.

68 C. T Dyrness, J. F. Franklin, and W. H. Moir, *A Preliminary Classification of Forest Communities in the Central Portion of the Western Oregon Cascades*, Bulletin No. 4, Coniferous Forest Biome, Ecosystem Analysis Studies, US/IBP: Forestry Sciences Laboratory, USDA Forest Service, Corvallis, Oregon, ii, 1–4. See Publications in https://andrewsforest.oregonstate.edu/; and Glenn M. Hawk, Jerry F. Franklin, W. Arthur McKee, and Randall B. Brown, *H. J. Andrews Experimental Forest Reference Stand System: Establishment and Use History*, Bulletin No. 12, Coniferous Forest Biome, Ecosystem Analysis Studies, US/IBP, 1–4, copy in SCARC.

69 Frederick A. Bierlmaier and Arthur McKee, *Climatic Summaries and Documentation for the Primary Meteorological Station, H. J. Andrews Experimental Forest, 1972 to 1984*, USDA Forest Service, Pacific Northwest Research Station, General Technical Report, PNW-GTR-242 (June 1989), 2–5.

70 Email communications from Mark Schulze and Fred Swanson, November 15, 2017.

71 Max G. Geier, "Jerry F. Franklin (1936-)," *Oregon Encyclopedia*, https://oregonencyclopedia.org/articles/franklin_jerry_f/; Coleman, *Big Ecology*, 42–48; and Geier interview with Bob Tarrant, July 24, 1997.

72 Max Geier, interviews with the H. J. Andrews IBP Group, February 10, 1998.

73 Geier, interviews with the H. J. Andrews IBP Group.

74 Geier, interviews with the H. J. Andrews IBP Group.

75 Golley, *History of the Ecosystem Concept*, 139–141.

76 This assessment is based on the summary of the NAS report published in *Science* 187 (February 21, 1975): 663.

77 Jerry Franklin, Retirement Lecture, November 22, 1991, Forest Science Laboratory, Corvallis, Oregon, in SCARC.

CHAPTER 2

1 R. H. Waring and J. F. Franklin, "Evergreen Coniferous Forests of the Pacific Northwest," *Science* (new series) 204, no. 4400 (June 1979): 1380–1386.

2 Charles Goodrich, Kathleen Dean Moore, and Frederick J. Swanson, eds. *In the Blast Zone: Catastrophe and Renewal on Mount St. Helens* (Corvallis: Oregon State University Press, 2008).

3 Glenn M. Hawk, Jerry F. Franklin, W. Arthur McKee, and Randall B. Brown, "H. J. Andrews Experimental Forest Reference Stand System: Establishment and Use History," Bulletin No. 12, *Ecosystem Analysis Studies*, US/IBP (1978), 1–3.

4 Hawk et al., "H. J. Andrews Experimental Forest Reference Stand System," 4; Lee E. Frelich, Meredith W. Cornett, and Mark A. White, "Controls and Reference Conditions in Forestry: The

Role of Old-Growth and Retrospective Studies," *Journal of Forestry* (Oct/Nov 2005): 439–344, and Stephen T. Jackson, "Humboldt for the Anthropocene," *Science* 365 (September 13, 2019): 1074–1076.

5 Richard Waring, "Internal Report 130, Contributions to a Better Management of Forest Ecosystems" (1973), 1–5, https://ir.library.oregonstate.edu/concern/defaults/vm40xs882.

6 R. H. Waring, "Structure and Function of the Coniferous Forest Biome Organization," in *Integrated Research in the Coniferous Forest Biome*, Bulletin 5, edited by R. H. Waring and R. L. Edmonds (University of Washington, 1974), 1–6, https://ir.library.oregonstate.edu/concern/defaults/th83m059w.

7 R. H. Waring and J. F. Franklin, "Evergreen Coniferous Forests of the Pacific Northwest," *Science* 204 (June 1979): 1380–1381.

8 Waring and Franklin, "Evergreen Coniferous Forests," 1382–1385.

9 Jerry F. Franklin and Richard H. Waring, "Distinctive Features of the Northwestern Coniferous Forest Development, Structure, and Function," in *Proceedings of the 40th Annual Biology Colloquium*, edited by Richard Waring (Corvallis: Oregon State University Press, 1980), 59–60, 64, 71–80.

10 Michel Batisse, "The Biosphere Reserve: A Tool for Environmental Conservation and Management," *Environmental Conservation* 9, no. 2 (Summer 1982): 101–102, 109; and Jennifer M. Thomsen, "An Investigation of the Critical Events and Influential Factors in the Evolution of the U.S. Man and the Biosphere Program," *Environmental Management* 61, no. 1 (2018): 1–4, https://link-springer-com.ezproxy.proxy.library.oregonstate.edu/article/10.1007/s00267-017-0988-z.

11 M. B. Dickerman (US Forest Service, Washington Office) to Regional Foresters, Directors, Area Directors, List of Biosphere Reserves in the United States, October 1, 1974; Report of the Task Force on Criteria and Guidelines for the Choice and Establishment of Biosphere Reserves, UNESCO, International Coordinating Council of the Program on Man and the Biosphere, Washington, DC, September 17–19, both in SCARC; and Jerry F. Franklin, "The Biosphere Reserve Program in the United States," *Science* 195 (January 1977): 262–267.

12 Franklin, "Biosphere Reserve Program," 262–266; and "UNESCO's Man and the Biosphere Program: What Are Biosphere Reserves All About?" George Wright Society, http://www.georgewright.org/mab, accessed May 10, 2017.

13 M. I. Dyer and M. M. Holand, "UNESCO's Man and the Biosphere Program," *BioScience* 38, no. 9 (1988): 635–641: and Andrew P. Vayda, "Progressive Contextualization: Methods for Research in Human Ecology," *Human Ecology* 11, no. 3 (1983): 2–3, 7.

14 Dyer and Holland, "UNESCO's Man and the Biosphere Program," 636–638.

15 Thomsen, "An Investigation," 3–4, 8–12.

16 Thomsen, "An Investigation," 13–15.

17 Mark Schulze to the author, February 23, 2018, and Thomsen, "An Investigation," 4, both in SCARC.

18 Schulze to the author. The Willamette National Forest withdrew from the International Network of Biosphere Reserves in 2017. See Suzanne Schindler and Paul Anderson to Paul Mungai, March 14, 2017, copies from Michael Nelson to the author in SCARC.

19 Charles F. Wilkinson, "The National Forest Management Act: The Twenty Years Behind, The Twenty Years Ahead," *University of Colorado Law Review* 68 (1997): 659–664; and Arnold W. Bolle, "The Bitterroot Revisited: A University Re-View of the Forest Service," *Public Land and Resources Review of the Forest Service* 1 (1989): 1–18.

20 Wilkinson, "National Forest Management Act," 664–665. Also see Jim Furnish, *Toward A Natural Forest: The Forest Service in Transition* (Corvallis: Oregon State University Press, 2015), 188.

21 Wilkinson, "National Forest Management Act," 671–676; and Kevin Marsh, *Drawing Lines in the Forest: Creating Wilderness Areas in the Pacific Northwest* (Seattle: University of Washington Press, 2007), 147.

22 Max Geier, interview with Art McKee, September 12, 1996.

23 Biological Research Program, Support of Field Research Facilities, NSF, Washington, DC (1975), SCARC.

24 James T. Callahan, "Long-Term Ecological Research," *BioScience* 34, no. 6 (June 1984): 363–364.

25 Author Unknown, "Preserving Sites for Long-Term Environmental Research," *Mosaic Magazine* (Jan/Feb 1976): 29–31. *Mosaic* was an NSF publication, in SCARC.

26 J. F. Franklin, "Scientific Reserves in the Pacific Northwest and Their Significance for Ecological Research," in *Proceedings of the Symposium on Terrestrial and Aquatic Ecological Studies of the Northwest, March 26–27, 1976* (Cheney: Eastern Washington State College, 1976), 201.

27 Jerry Franklin to William Ferrell et al., Preparing a NSF Proposal for Support of H. J. Andrews Experimental Forest, December 8, 1975, USDA Forest Service, Forest Science Laboratory, Corvallis, OR, in SCARC.

28 H. J. Andrews Experimental Ecological Reserve Proposal (1976), 1–2, in SCARC.

29 H. J. Andrews Experimental Ecological Reserve Proposal (1976), 3–8.

30 *Long-Term Ecological Measurements: Report of a Conference*, Woods Hole, Massachusetts (March 16–18, 1977), 1, in SCARC.

31 *Long-Term Ecological Measurements*, 2–3, 5, 9, 19.

32 *Experimental Ecological Reserves* (Washington, DC: GPO, June 1977), 4–9, 15, 21, 29.

33 Jerry F. Franklin to Stanley Cook et al., invitations to join the national advisory committee, July 25, 1977, in SCARC.

34 Research Proposal submitted to the National Science Foundation for Second Year Support of Physical and Biological Processes in the Response of a Coniferous Forest Ecosystem to Perturbation (1978), in SCARC. For a sampling of chronosequence studies, see Peter Schoonmaker and Arthur McKee, "Species Composition and Diversity during Secondary Succession of Coniferous Forests in the Western Cascade Mountains of Oregon," *Forest Science* 34, no. 4 (December 1988): 960–979; and Mark E. Harmon, *Long-Term Experiments on Log Decomposition at the H. J. Andrews Experimental Forest*, USDA Forest Service, Pacific Northwest Research Station, General Technical Report, PNW-GTR-280 (January 1992).

35 George Lauff and David Reichle, "Experimental Ecological Reserves," *Bulletin of the Ecological Society of America* 60, no. 1 (March 1979): 4. The Lauff and Reichle article lists seventy-one sites, including the H. J. Andrews Experimental Forest and many Research Natural Areas.

36 Callahan, "Long-Term Ecological Research," 364.

37 Report of the National Advisory Committee of the H. J. A. with Funding Power-Brokers such as the National Science Foundation. Andrews EER Site Visit (September 19–21, 1978), 1–4, in SCARC.

38 Report of the National Advisory Committee (1978), 7–9.

39 Proposed Supplement to Grant for Support of the H. J. Andrews Experimental Forest as a National Field Research Facility, DEB-76-11978, principal investigator, Richard Waring, August 25, 1978, in SCARC.

40 Jerry Franklin and Arthur McKee, Facilities Development Plan for H. J. Andrews Experimental Forest and Associated Research Natural Areas, Willamette National Forest and Pacific Norwest Forest and Range Experiment Station (October 15, 1979), 1–2, in SCARC.

41 Franklin and McKee, Facilities Development Plan, 3–5.

42 Franklin and McKee, Facilities Development Plan, 11.

43 Franklin and McKee, Facilities Development Plan, 12.

44 Charles M. Crisafulli, Frederick J. Swanson, and Virginia H. Dale, "Overview of Ecological Responses to the Eruption of Mount St. Helens: 1980–2005," in *Ecological Responses to the 1980 Eruption of Mount St. Helens*, edited by Dale, Swanson, and Crisafulli (Springer Science, 2005), 287; and Swanson to the author, May 7, 2019.

45 Scott Slovic, "Foreword," in *In the Blast Zone: Catastrophe and Renewal on Mount St. Helens*, edited by Charles Goodrich, Kathleen Dean Moore, and Frederick J. Swanson (Corvallis: Oregon State University Press, 2008), vii–viii; Jerry F. Franklin, "Foreword," *Ecological Responses*, v–vii; and *Oregonian*, May, 13, 1990.

46 *Oregonian*, May 13, 1990.

47 *Oregonian*, May 13, 1990.

48 Dale et al., "Preface," *Ecological Responses*, ix–xi.

49 Crisafulli et al., "Overview of Ecological Responses," *Ecological Responses*, 287–288.

50 Crisafulli et al., "Overview of Ecological Responses," *Ecological Responses*, 298.

51 Max Geier, interview with Jerry Franklin, September 13, 1996; and Geier interview with Fred Swanson, September 6, 1996.

52 Geier interview with Franklin, September 13, 1996.

53 Geier interview with Franklin, September 13, 1996.

54 Geier interview with Swanson, September 6, 1996; Geier interview with Gordon Grant, October 10, 1997; and Simmons B. Buntin, "Dirty Words on Mount St. Helens," *Terrain.org: A Journal*

of the Built and Natural Environments 26 (Fall/Winter 2010), http://www.terrain.org/columns/26/buntin.htm, np. For the 2010 pulse, also see Elizabeth Dodd, "Isogloss: Language and Legacy on Mount St. Helens," *Georgia Review* (Spring 2012): 104–115, https://andrewsforest.oregonstate.edu/sites/default/files/lter/pubs/pdf/pub4762.pdf.

55 Geier interview with Swanson, September 6, 1996.

56 Geier interview with Jim Sedell, February 14, 1998; Geier interview with the Riparian Group, November 21, 1997; and Geier interview with Art McKee, September 12, 1996.

57 NSF, "A New Emphasis in Long-Term Research," December 25, 1979, in SCARC. The first three long-term research award cycles were for five years. Beginning in 1996 and continuing, the funding awards were for six-year cycles. Lina DiGregorio to the author, April 23, 2019. See appendix A for a listing of the LTER award periods and the principal investigator for each.

58 NSF, "A New Emphasis."

59 Research Agreement, Operation of the H. J. Andrews Experimental Forest, Supplement No. PNW-80-256, March 17, 1980, to March 17, 1990, authorized principal investigator, Richard H. Waring, in SCARC.

60 Long-Term Ecological Research on the Andrews Experimental Forest, Proposal to the National Science Foundation, Division of Environmental Biology, February 4, 1980, Richard Waring, principal investigator, cover pp. and p. 19, in SCARC.

61 Long-Term Ecological Research on the Andrews Experimental Forest, 1–5.

62 Long-Term Ecological Research on the Andrews Experimental Forest, 16–18, 59–60.

63 J. T. Callahan to Richard H. Waring, May 6, 1980; and Waring to Callahan, May 23, 1980, both in SCARC. Callahan, "Long-Term Ecological Research," 365.

64 Long-Term Research on the Andrews Experimental Forest, Richard Waring, principal investigator, Proposal to the National Science Foundation, April 24, 1981, in SCARC.

65 National Advisory Committee Report (April 1981), 1–2, in SCARC.

66 National Advisory Committee Report (April 1981), 3.

67 Proposal to the National Science Foundation, Ecosystems Studies Program, Division of Environmental Biology, "Long-Term Ecological Research on the Andrews Experimental Forest," Jerry Franklin, principal investigator, Continuing Award Request (August 26, 1982), 1–4, in SCARC.

68 Proposal to the National Science Foundation (August 26, 1982), 6–7.

69 Progress Report for Year Three Long-Term Ecological Research Program, H. J. Andrews Experimental Forest; and Long-Term Ecological Research (II) on the H. J. Andrews Experimental Forest, Proposal to the National Science Foundation, Fred Swanson, principal investigator, August 21, 1986, both in SCARC.

70 John E. Hobbie, Stephen R. Carpenter, Nancy B. Grimm, James R. Grosz, and Timothy R. Seastedt, "The US Long Term Ecological Research Program," *BioScience* 53, no. 1 (January 2003), 21, 25, 28, 29.

71 Frederick J. Swanson and Richard E. Sparks, "Long-Term Ecological Research and the Invisible Place," *BioScience* 40, no. 7 (Jul/Aug, 1990): 502–504.

72 *Oregonian*, December 4, 1985, and July 23, 1987.

73 R. H. Waring, "Structure and Function of the Coniferous Forest Biome Organization," in "Integrated Research in the Coniferous Forest Biome," *Coniferous Forest Biome Ecosystem Analysis Studies*, Bulletin No. 5 (September 1974), 1 and 4; and the author's conversations with Richard Waring, December 11, 2017, in SCARC.

74 Arthur McKee, "Integrating Academic and Agency Research at the H. J. Andrews Experimental Forest," *Research Natural Areas: Baseline Monitoring and Management*, USDA Forest Service, Intermountain Forest and Range Experiment Station, Ogden, Utah, General Technical Report, INT-173 (November 1984), 54–58.

75 Paul W. Hirt, *A Conspiracy of Optimism: Management of the National Forests since World War Two* (Lincoln: University of Nebraska Press, 1994), 272; William G. Robbins, *Landscapes of Conflict: The Oregon Story, 1940–2000* (Seattle: University of Washington Press, 2004), 205; and Robbins, *Hard Times in Paradise: Coos Bay, Oregon*, rev. ed. (Seattle: University of Washington Press, 2006), 166–180.

76 Catherine Caulfield, "The Ancient Forest," *New Yorker*, May 14, 1990, 82–83; and Robbins, *Landscapes of Conflict*, 205–206.

CHAPTER 3

1   Jacob Bendix and Carol M. Liebler, "Place, Distance, and Environmental News: Geographic Variation in Newspaper Coverage of the Spotted Owl Conflict, *Annals of the Association of American Geographers* 89, no. 4 (1999): 662.

2   Samuel Hays, *War in the Woods: The Rise of Ecological Forestry in America* (Pittsburgh: University of Pittsburgh Press, 2007); and Douglas Bevington, *The Rebirth of Environmentalism: Grassroots Activism from the Spotted Owl to the Polar Bear* (Washington, DC: Island Press, 2009).

3   Craig Hill, "Washington's 31 Wilderness Areas," Tacoma *News-Tribune*, August 31, 2014, http://www.thenewstribune.com/outdoors/article25878187.html; Wilderness.net, *Wilderness Areas in Oregon*, https://www.wilderness.net/NWPS/stateView?state=OR; and J. F. Franklin and R. H. Waring, "Information on the H. J. Andrews Experimental Forest, Western Oregon Cascade Range," in *Coupling of Ecological Studies with Remote Sensing: Potentials at Four Biosphere Reserves in the United States*, edited by M. I. Dyer and D. A. Crossley, Pub. 9504 (Washington, DC: Department of State, Bureau of Oceans and International Environmental and Scientific Affairs, 1986), 37–38.

4   Franklin and Waring, "Information," 39–40.

5   Franklin and Waring, "Information," 41.

6   Max Geier, interview with Mike Kerrick, August 28, 1996; and Jay Gashwiler, "Plant and Mammal Changes on a Clearcut in West-Central Oregon," *Ecology* 51, no. 6 (Autumn 1970): 1018–1026.

7   Larry D. Harris, Chris Maser, and Arthur McKee, "Patterns of Old Growth Harvest and Implications for Cascades Wildlife," *Transactions of the 47th North American Wildlife and Natural Resources Conference* (Washington, DC: Wildlife Management Institute, 1982), 374–384.

8   Jerry Franklin Retirement Lecture, Forest Sciences Laboratory, Oregon State University, November 22, 1991, in SCARC; and Jerry F. Franklin, Kermit Cromack Jr., William Denison, Arthur McKee, Chris Maser, James Sedell, Fred Swanson, and Glen Juday, *Ecological Characteristics of Old-Growth Douglas-Fir Forests*, USDA Forest Service, Pacific Northwest Forest and Range Experiment Station, General Technical Report PNW-118 (February 1981), summary, np.

9   Franklin et al., *Ecological Characteristics of Old-Growth Douglas-Fir Forests*, abstract, 1–2, 42; and Jon R. Luoma, "Untidy Wonder," *Discover* (October 1992), 93.

10 Eric Forsman, "A Preliminary Investigation of the Spotted Owl in Oregon" (MS thesis, Oregon State University, 1976), unpaginated abstract, and 1–2; and Eric D. Forsman, E. Charles Meslow, and Monica J. Strub, "Spotted Owl Abundance in Young versus Old-Growth Forests, Oregon," *Wildlife Society Bulletin* 5, no. 2 (Summer 1977): 43–47, in SCARC.

11 Forsman, "Preliminary Investigation of the Spotted Owl," 86–88.

12 Erik Forsman, "Habitat Utilization by Spotted Owls in the West-Central Cascades of Oregon" (PhD diss., Oregon State University, 1980), unpaginated abstract.

13 Forsman, "Habitat Utilization," unpaginated abstract, and 79–83.

14 Eric D. Forsman, Kirk M. Horn, and William A. Neitro, "Spotted Owl Research and Management," *USDA Forest Service / UNL Faculty Publications* (1982) Digital Commons, University of Nebraska-Lincoln, http://digitalcommons.unl.edu/usdafsfacpub/73.

15 Eric Forsman, E. Charles Meslow, and Howard M. Wight, *The Distribution of the Spotted Owl in Oregon* (Corvallis: Oregon Agricultural Experiment Station Technical Paper 6251, 1984), 7–10.

16 Forsman et al., *Distribution*, 16, 53–56.

17 Miles Burnett and Charles Davis, "Getting Out the Cut: Politics and National Forest Timber Harvests, 1960–1995," *Administration and Society* (2002): 2–9, 211, https://doi.org/10.1177/0095399702034002004.

18 Proposal to the National Science Foundation, Ecosystems Studies Program, "Long-Term Ecological Research (II) on the Andrews Experimental Forest," Jerry Franklin, principal investigator (April 9, 1985), 2, in SCARC. Appended to the Andrews application was a lengthy appeal for upgrading the site's computer equipment, which was "wearing out and obsolete." The point to the request was that the Andrews' site "received the first and therefore the oldest of the current equipment." See LTERS Year Computer Plan, in SCARC.

19 A. J. Hansen, T. A. Spies, F. J. Swanson, and J. L. Ohmann, "Conserving Biodiversity in Managed Forests," *BioScience* 41, no. 6 (1991): 91–108; and Jack Ward Thomas, "Needs for and Approaches to Wildlife Habitat Assessment," *USDA Forest Service / UNL Faculty Publications* (1982), http://digitalcommons.unl.edu/usdafsfacpub/81, pp. 35–41.

20 Thomas, "Needs for and Approaches to," 41–44; and Burnett and Davis, "Getting Out the Cut," 209.

21 Jerry F. Franklin and Thomas A. Spies, "Characteristics of Old-Growth Douglas-Fir Forests,"
    reprinted from *New Forests for a Changing World: Proceedings of the 1983 Society of American Forests
    National Convention* (Portland: Society of American Foresters, 1983), unpaginated, https://an-
    drewsforest.oregonstate.edu/sites/default/files/lter/pubs/pdf/pub120.pdf; and Jon R. Luoma,
    "A Wealth of Forest Species Is Found Underfoot," *New York Times*, July 2, 1991.

22 Franklin and Spies, "Characteristics of Old-Growth Douglas-Fir Forests."

23 Arthur McKee, "Integrating Academic and Agency Research Interests at the H. J. Andrews Ex-
    perimental Forest," *Research Natural Areas: Baseline Monitoring and Management, Proceedings of a
    Symposium in Missoula, Montana, March 24, 1984*, Janet L. Johnson, Jerry F. Franklin, and Richard
    G. Krebill, coordinators, USDA Forest Service, Intermountain Forest and Range Experiment
    Station, Ogden, Utah, General Technical Report, Int-173 (November 1984), 54.

24 McKee, "Integrating," 56.

25 McKee, "Integrating," 56 and 58.

26 Proposal to the National Science Foundation, Ecosystems Studies Program, Division of Biotic
    Systems and Resources, "Long-Term Ecological Research on the Andrews Experimental Forest,"
    Jerry Franklin, principal investigator (September 9, 1983), 5–6, in SCARC.

27 Proposal to the National Science Foundation (September 9, 1983), 22; and Proposal to the Na-
    tional Science Foundation, Ecosystems Studies Program, "Long-Term Ecological Research (II)
    on the Andrews Experimental Forest," Jerry Franklin, principal investigator (August 20, 1986),
    18–19, in SCARC.

28 Proposal to the National Science Foundation, Ecosystems Studies Program, "Long-Term Eco-
    logical Research (II) on the Andrews Experimental Forest," Fred Swanson, principal investigator
    (September 6, 1987), 3–4, 13–14, in SCARC.

29 LTER annual reports for 1984, 1986, and 1987, in SCARC.

30 Proposal to the National Science Foundation, Ecosystems Studies Program, "Long-Term Eco-
    logical Research (II) on the Andrews Experimental Forest (year 5)," Frederick J. Swanson, prin-
    cipal investigator (December 1989), 3–7, in SCARC.

31 Proposal to the National Science Foundation (December 1989), 16.

32 Proposal to the National Science Foundation (December 1989), 19.

33 Proposal to the National Science Foundation (December 1989), 19–20.

34 Proposal to the National Science Foundation, "Long-Term Ecological Research on the Andrews
    Experimental Forest," Frederick J. Swanson, principal investigator, January 26, 1990, in SCARC.

35 James E. Schindler to Frederick J. Swanson, January 9, 1991; and Panel Summary, March 15,
    1990, all in NSF packet to John V. Byrne, Dean of Research, Oregon State University, LTER3
    Award, in SCARC.

36 Some of this argument reflects my reading of David J. Brooks and Gordon E. Grant's *New Perspec-
    tives in Forest Management: Background, Science Issues, and Research Agendas*, USDA Forest Service,
    Pacific Northwest Research Station, PNW-RP-456 (September 1992), 1–14.

37 Jerry F. Franklin, Thomas Spies, David Perry, Mark Harmon, and Arthur McKee, "Modifying
    Douglas-Fir Management Regimes for Nontimber Objectives," in *Douglas-Fir: Stand Manage-
    ment for the Future: Proceedings of a Symposium*, edited by Chadwick Oliver, Donald Hanley, and
    Jay Johnson (Seattle: College of Forest Resources, University of Washington, 1986), 373–379;
    and Jerry Franklin and Richard T. T. Forman, "Creating Landscape Patterns by Forest Cutting:
    Ecological Consequences and Principles," *Landscape Ecology* 1, no. 1 (1987): 5–18.

38 Jerry F. Franklin, "The 'New Forestry,'" *Journal of Soil and Water Conservation* (Nov/Dec 1989):
    549; J. F. Franklin, "Old Growth: The Contribution to Commercial Forests," *Forest Planning
    Canada* 5, no. 3 (1989): 17–23; and *Oregonian*, August 28, 1989.

39 Anna Marie Gillis, "The New Forestry: An Ecosystem Approach to Land Management," *BioSci-
    ence* 40, no. 8 (September 1990): 558–562.

40 Jack Ward Thomas et al., "A Conservation Strategy for the Northern Spotted Owl: Report of the
    Interagency Scientific Committee to Address the Conservation of the Northern Spotted Owl,"
    unpublished interagency document (Portland, 1990), 1–2, 7, 14, in https:/Section 318/scholar.
    google.com/scholar?hl=en&as_sdt=0%2C38&q=Interagency+Scientific+Committee+for+the
    +Northern+Spotted+Owl&btnG=.

41 Thomas et al., "A Conservation Strategy," 8.

42 Thomas et al., "A Conservation Strategy," 2–4, 7.

43 Bruce G. Marcot and Jack Ward Thomas, *Of Spotted Owls, Old Growth, and New Policies: A History
    Since the Interagency Scientific Report*, USDA Forest Service, Pacific Northwest Research Station,
    General Technical Report, PNW GTR-408 (Portland, 1997), 1–4; and Mark Bonnett and Kurt

Zimmerman, "Politics and Preservation: The Endangered Species Act and the Northern Spotted Owl," *Ecology Law Quarterly* 18, no. 1 (January 1991): 124–128. The agreement to informally protect the owl included the creation of the Interagency Scientific Committee.

44 Michael C. Blumm, "Ancient Forests and the Supreme Court: Issuing a Blank Check for Appropriation Riders," *Washington University Journal of Urban and Contemporary Law* 43 (January 1993): 40–44; Marcot and Thomas, *Of Spotted Owls,* 2–4; Burnett and Davis, "Getting Out the Cut," 210–212; and John Klein Robbehaar, "Judicial Review of Forest Service Timber Sales: Environmental Plaintiffs Gain New Options under the Oregon Wilderness Act," *Natural Resources Journal* 35 (Winter 1995): 209–215.

45 National Research Council, Committee of Forestry Research, *Forestry: A Mandate for Change* (Washington, DC: National Academy Press, 1990), v–vi, 1–2.

46 National Research Council, *Forestry,* 54–58.

47 *New York Times,* March 3, 1990; and Richard W. Behan, *Plundered Province: Capitalism, Politics, and the Fate of the Federal Lands* (Washington, DC: Island Press, 2001), 35.

48 *New York Times,* October 4, 1991.

49 Bill Dietrich, "Old-Growth Mosaic," *Seattle Times,* October 2, 1989.

50 Michael Oreskes, "Anti-Incumbent Fever Threatens Oregon Senator," *New York Times,* October 22, 1990.

51 *New York Times,* September 20 and 28, 1992.

52 Corvallis *Gazette-Times,* February 9, 1990; and Eugene *Register-Guard,* June 23, 1990.

53 Eugene *Register-Guard,* June 23, 1990.

54 Salem *Statesman Journal,* June 2, 1991.

55 Hal Salwasser, "New Perspectives for Sustaining Diversity in the U.S. National Forest System," *Conservation Biology* 5, no. 4 (1991): 567–569; Jon R. Luoma, "New Government Plan for National Forests Generates a Debate," *New York Times,* June 30, 1992; *New York Times,* June 13, 1990; and J. Stan Rowe, "A New Paradigm for Forestry," *Forestry Chronicle* 70, no. 5 (Sept/Oct 1994): 565–568.

56 William Atkinson, "Another View of New Forestry," presented to the Oregon Chapter of the Society of American Foresters, May 4, 1990, Eugene, H. J. Andrews Experimental Forest, Publications, http://andlter.forestry.oregonstate.edu/data/Bibilo/Bibliography.aspx.

57 *Oregonian,* October 15, 1990.

58 Charles W. Philpot to George Brown, April 7, 1992; and Philpot to Jennifer O'Loughlin, April 14, 1992, both in SCARC.

59 Dean S. DeBell, "Silviculture and New Forestry in the Pacific Northwest," *Journal of Forestry* 91, no. 12 (1993): 26–30.

60 F. J. Swanson and J. F. Franklin, "New Forestry Principles from Ecosystem Analysis of Pacific Northwest Forests," *Ecological Applications* 2, no. 3 (August 1992): 262–274.

CHAPTER 4

1 Mark Bonnett and Kurt Zimmerman, "Politics and Preservation: The Endangered Species Act and the Northern Spotted Owl," *Ecology Law Quarterly* 18, no. 1 (1991): 107.

2 Andy Kerr, "Starting the Fight, Finishing the Job," 129–135; and Jack Ward Thomas, "Increasing Difficulty of Active Management on National Forests—Problems and Solutions," 189–193, both in *Old Growth in a New World: A Pacific Northwest Icon Reexamined,* edited by Thomas A. Spies and Sally L. Duncan (Washington, DC: Island Press, 2009).

3 Bonnett and Zimmerman, "Politics and Preservation," 168–171.

4 Johnson, K. Norman. *Alternatives for Management of Late-Successional Forests of the Pacific Northwest: A Report to the Agriculture Committee and the Merchant Marine and Fisheries Committee of the US House of Representatives* (Washington, DC: GPO, 1991); and E. Charles Meslow, "Spotted Owl Protection: Unintentional Evolution toward Ecosystem Management," *Endangered Species UPDATE: Exploring an Ecosystem Approach to Endangered Species* 10, no. 3/4 (1993): 34–37.

5 Meslow, "Spotted Owl Protection," 37–38; and Thomas, "Increasing Difficulty of Active Management," 193.

6 Jerry F. Franklin, "Preserving Biodiversity: Species, Ecosystems, or Landscapes?" *Ecological Applications* 3, no. 2 (May 1993): 202–205.

7 Frederick J. Swanson, Julia A. Jones, and Gordon E. Grant, "The Physical Environment as a Basis for Managing Ecosystems," in *Creating a Forestry for the 21st Century: The Science of Ecosystem*

Management, edited by Kathryn A. Kohm and Jerry F. Franklin (Washington, DC: Island Press 1997), 229–237.

8 *New York Times,* September 20, 1992.

9 *New York Times,* September 20, 1992. Strong third-party candidate Ross Perot joined Bush in proposing revisions to the Endangered Species Act. See Thomas, "Increasing Difficulty of Active Management," 194.

10 *Oregonian,* March 27, 1993.

11 William J. Clinton, "Remarks on Opening the Forest Conference in Portland, Oregon," April 2, 1993, The American Presidency Project, University of California, Santa Barbara, http://www.presidency.ucsb.edu/ws/?pid=46396.

12 *Forest Ecosystem Management: An Ecological, Economic, and Social Assessment, Report of the Forest Ecosystem Management Assessment Team* (Portland, July 1993), v–xi; and Paul W. Hirt, *A Conspiracy of Optimism: Management of the National Forests since World War Two* (Lincoln: University of Nebraska Press, 1994), 288–292.

13 *Forest Ecosystem Management,* in William G. Robbins Papers, Special Collections and Archives, Oregon State University (hereafter SCARC).

14 Jerry F. Franklin, "Scientists in Wonderland," *BioScience Supplement* (1995), S-78; and Fred Swanson to the author, May 5, 2018.

15 Marcot and Thomas, *Of Spotted Owls,* 4–5; Franklin, "Scientists in Wonderland," S-77; and Jack Ward Thomas, Jerry F. Franklin, and K. Norman Johnson, "The Northwest Forest Plan: Origins, Components, Implementation Experience, and Suggestions for Change," *Conservation Biology* 20, no. 2 (2006): 277–287.

16 "The Clinton Forest Plan," Institute for Social Ecology, http://social-ecology.org/wp/1994/04/the-clinton-forest-plan/.

17 Thomas et al., "Northwest Forest Plan," 281.

18 I am indebted to Thomas Spies for this summary of the Northwest Forest Plan. See Thomas Spies, Box 2.1., "The Northwest Forest Plan for Federal Lands," in Spies and Duncan, *Old Growth in a New World,* 21; and Jim Furnish, *Toward A Natural Forest: The Forest Service in Transition* (Corvallis: Oregon State University Press, 2015), 119.

19 Thomas et al., "Northwest Forest Plan," 281; and Randy Molina, Bruce G. Marcot, and Robin Lesher, "Protecting Rare, Old-Growth Associated Species under the Survey and Manage Program: Guidelines of the Northwest Forest Plan," *Conservation Biology* 20, no. 2 (2006): 307.

20 Thomas et al., "Northwest Forest Plan," 282–283.

21 Molina et al., "Protecting Rare, Old-Growth Associated Species," 310–311.

22 Molina et al., "Protecting Rare, Old-Growth Associated Species," 315; and Thomas et al., "Northwest Forest Plan," 284.

23 Molina et al., "Protecting Rare, Old-Growth Associated Species," 315–317.

24 Thomas et al., "Northwest Forest Plan," 285–286.

25 Nancy M. Diaz, "Northwest Forest Plan," *Proceedings from the Ridge—Considerations for Planning at the Landscape Scale,* compiled by Hermann Gucinski, Cynthia Minor, and Becky Bittner, USDA Forest Service, Pacific Northwest Research Station, General Technical Report, PNW-GTR-596 (January 2004), 87–91.

26 Raymond J. Davis et al., *Status and Trends of Late-Successional and Old-Growth Forests,* USDA Forest Service, Pacific Northwest Research Station, General Technical Report, PNW-GTR-911 (December 2015), executive summary, and 2–10.

27 Davis et al., *Status and Trends,* 109–110.

28 Richard W. Haynes et al., *Northwest Forest Plan: The First 10 Years (1994–2003): Synthesis of Monitoring and Research Results,* USDA Forest Service, Pacific Northwest Research Station, Portland, Oregon, General Technical Report, PNW-GTR-651 (October 2006), vii–xi.

29 Haynes et al., *Northwest Forest Plan,* xii–xiii.

30 Haynes et al., *Northwest Forest Plan,* xiv–xvii; and K. Norman Johnson and Frederick J. Swanson, "Historical Context of Old-Growth Forests in the Pacific Northwest: Policy, Practice, and Competing World Views," in *Old Growth in a New World: A Pacific Northwest Icon Reexamined,* edited by Thomas A. Spies and Sally L. Duncan (Washington, DC: Island Press, 2009), 19–23.

31 Susan Charnley, "The Northwest Forest Plan as a Model for Broad-Scale Ecosystem Management: A Social Perspective," *Conservation Biology* 20, no. 2 (2006): 330–332; William G. Robbins, *Oregon, This Storied Land* (Portland: Oregon Historical Society Press, 2005), 179–181; and Robbins, *Hard Times in Paradise, Coos Bay, Oregon,* rev. ed. (Seattle: University of Washington

Press, 2006), 189–192. Jim Furnish, supervisor of the Siuslaw National Forest at the time, indicates that harvests on the Siuslaw dropped from 215 to 23 million board feet, a 92 percent reduction. See Furnish, *Toward A Natural Forest*, 119.

32 Charnley, "Northwest Forest Plan," 232–234.

33 Charnley, "Northwest Forest Plan," 336–338.

34 Thomas et al., "Northwest Forest Plan," 277.

35 C. S. Holling, *Adaptive Environmental Assessment and Management* (London: John Wiley and Sons, 1978); Carl Walters, *Adaptive Management of Renewable Resources* (New York: Macmillan, 1986), vii–viii, 3–4; Kai N. Lee, *Compass and Gyroscope: Integrating Science and Politics for the Environment* (Washington, DC: Island Press, 1993), 65; and George Stankey and Bruce Shindler, *Adaptive Management Areas: Achieving the Promise, Avoiding the Peril*, USDA Forest Service, Pacific Northwest Research Station, PNW-GTR-394 (March 1997), 1–2.

36 Bernard T. Bormann, Richard W. Haynes, and Jon R. Martin, "Adaptive Management of Forest Ecosystems: Did Some Rubber Hit the Road?" *BioScience* 57, no. 2 (February 2007): 186–187; Bruce Shindler, Brent Steel, and Peter List, "Public Judgments of Adaptive Management," *Journal of Forestry* 94, no. 6 (June 1996): 4–5; George H. Stankey et al., "Adaptive Management and the Northwest Forest Plan: Rhetoric and Reality," *Journal of Forestry* 101, no. 1 (Jan/Feb 2003): 40–41; and Forest Ecosystem Assessment Team, *Forest Ecosystem Management: An Ecological, Economic, and Social Assessment* (Portland: USDA Forest Service, 1993), iii.

37 Stankey and Shindler, *Adaptive Management Areas*, 1–9.

38 Stankey and Shindler, *Adaptive Management Areas*, 11–18. Swanson believes that although locals could criticize, they lacked experience in field applications. Swanson to the author, May 14, 2019.

39 J. H. Cissel, F. J. Swanson, W. A. McKee, and A. L. Burdett, "Using the Past to Plan the Future in the Pacific Northwest," *Journal of Forestry* 92, no. 8 (August 1994): 30–31, 46.

40 John H. Cissel et al., *A Landscape Plan Based on Historical Fire Regimes for a Managed Forest Ecosystem: The Augusta Creek Study*, USDA Forest Service, Pacific Northwest Research Station, General Technical Report, PNW-GTR-422 (May 1988), unpaginated abstract, and p. 5.

41 Cissel et al., *A Landscape Plan*, unpaginated summary, and p. 4.

42 Cissel et al., *A Landscape Plan*, 5–7.

43 Cissel et al., *A Landscape Plan*, 8.

44 Cissel et al., *A Landscape Plan*, 73–75.

45 Shindler et al., "Public Judgments of Adaptive Management," 4–6; and John Cissel, "Research and Learning Assessment for the Cascades Adaptive Management Area," prepared for the Cascades Adaptive Management Steering Team," report for the Cascade Center for Ecosystem Management (Oregon State University, 1995), 1, in SCARC.

46 Cissel, "Research and Learning Assessment," 1–6.

47 Blue River Landscape Project: Landscape Management and Monitoring Strategy (1997), Central Cascades Adaptive Management Area, Report on file at the Blue River Ranger District, Blue River, Oregon. The report is cited in Peter J. Weisberg, *An Evaluation of the Blue River Landscape Project: How Well Does It Use Historical Fire Regimes as a Model? Final Report to the US Forest Service, Blue River Ranger District, Blue River, Oregon* (February 1999). For this document and John Cissel and Fred Swanson, *Blue River Landscape Study: Testing an Alternative Approach*, October 13, 1999, see H. J. Andrews Publications, http://andlter.forestry.oregonstate.edu/data/Bibilo/Bibliography.aspx.

48 Cissel and Swanson, *Blue River Landscape Study*, 4–5; and Andrew Gray and Catherine Miller, "Vegetation Change in the Blue River Landscape Study: 1998–2005," report in H. J. Andrews Publications, http://andlter.forestry.oregonstate.edu/data/Bibilo/Bibliography.aspx.

49 John H. Cissel, Frederick J. Swanson, and Peter J. Weisberg, "Landscape Management Using Historical Fire Regimes: Blue River, Oregon," *Ecological Applications* 9, no. 4 (1999): 1217–1230. For an assessment of water temperature in Blue River headwater streams, see "Vital Signs: Water Temperature Monitoring in Headwaters Streams in the Blue River Landscape Study," *Cascade Center Research and Management News*, no. 6 (Winter 1998/1999): 2–3, in SCARC.

50 George H. Stankey, Bernard T. Bormann, and Roger N. Clark, eds., *Learning to Manage a Complex Ecosystem: Adaptive Management and the Northwest Forest Plan*, USDA Forest Service, Pacific Northwest Research Station, Research Paper, PNW-RP-567 (August 2006), unpaginated summary.

51 Stankey et al., *Learning to Manage*, summary.

52 Stankey et al., *Learning to Manage*, summary.

53 Stankey et al., "On Closing," in *Learning to Manage*, 177–180.

54 Fred Swanson, HG_ms_Frednotes_082418, notes on a coauthored paper about science/policy management/and the Andrews Forest, in SCARC.

55 Raymond J. Davis et al., *Status and Trends of Late Successional and Old-Growth Forests*, USDA Forest Service, Pacific Northwest Research Station, General Technical Report, PNW-GTR-911 (December 2015), abstract, and p. 1.

56 Davis et al., *Status and Trends*, 11–12, 47–48.

57 Davis et al., *Status and Trends*, 53–54; and "Northwest Forest Plan 20-Year Monitoring Draft Reports Released Today," Portland, June 8, 2015, https://wro.fs.fed.us/pnw/news/2015/06/northwest-forest-plan.shtml.

CHAPTER 5

1   Alston Chase, "The West Is Burning and Humans, Animals and the Eco-System Pays for the Green Policy," *SUAnews.com* (1994), https://www.suanews.com/alston-chase/the-west-is-burning-2.html.

2   The ideas presented here are gleaned from Kathryn A. Kohm and Jerry F. Franklin, eds., *Creating a Forestry for the 21st Century: The Science of Ecosystem Management* (Washington, DC: Island Press, 1993); see the following sections: Jack Ward Thomas, "Foreword," ix–xii; Kathryn Kohm and Jerry F. Franklin, "Introduction," 1–5; and David A. Perry and Michael P. Amaranthus, "Disturbance, Recovery, and Stability," 31–34.

3   Julian L. Simon, "Resources, Population, Environment: An Oversupply of False, Bad News," *Science*, new series 208, no. 4451 (June 27, 1980): 1431–1433.

4   Simon, "Resources, Population, Environment," 1434–1436.

5   R. Edward Grumbine, *Ghost Bears: Exploring the Biodiversity Crisis* (Washington, DC: Island Press, 1992), 155–157; and Alston Chase, *In a Dark Wood: The Fight over Forests and the Rising Tyranny of Ecology* (Boston: Houghton Mifflin, 1995), 7.

6   Frank Benjamin Golley, *The History of the Ecosystem Concept in Ecology: More Than the Sum of the Parts* (New Haven, CT: Yale University Press, 1993), 8; and David C. Coleman, *Big Ecology: The Emergence of Ecosystem Science* (Berkeley: University of California Press, 2010), 26–27.

7   Edmonds is quoted in Coleman, *Big Ecology*, 45; Max Geier IBP interview, February 10, 1998; and Geier interview with Jerry Franklin, September 13, 1996.

8   Geier interview with IBP group, February 10, 1998; Geier interview with Jerry Franklin, September 13, 1996, both in SCARC; and Coleman, *Big Ecology*, 17.

9   J. F. Franklin, "Past and Future of Ecosystem Research: Contribution of Dedicated Experimental Sites," in *Forest Hydrology and Ecology at Coweeta* (New York: Springer-Verlag, 1988), 415–417.

10 Franklin, "Past and Future," 418–423.

11 F. J. Swanson, R. P. Neilson, and G. E. Grant, "Some Emerging Issues in Watershed Management: Landscape Patterns, Species Conservation, and Climate Change," in *Watershed Management: Balancing Sustainability and Environmental Change*, edited by Robert J. Naiman (New York: Springer-Verlag, 1992), 307–323.

12 Merrill R. Kaufmann et al., *An Ecological Basis for Ecosystem Management*, USDA Forest Service, General Technical Report RM 246 (Fort Collins, 1994), 1–3.

13 David N. Bengston, "Changing Forest Values and Ecosystem Management," *Society and Natural Resources* 7 (1994): 515–518.

14 Grumbine, *Ghost Bears*, 22–23, 155–156.

15 Grumbine, *Ghost Bears*, 60–63, 236–237.

16 Jerry F. Franklin, "Lessons from Old-Growth," 11–13; and Bruce Lippke and Chadwick Oliver, "Managing for Multiple Values," 14–18, both in *Journal of Forestry* 91, no. 12 (1993).

17 Rachel Allison, "Administration Policies and Old-Growth: An Interview with Assistant Secretary of Agriculture, James R. Lyons," 20–23; and Dean S. DeBell and Robert O. Curtis, "Silviculture and New Forestry in the Pacific Northwest," 25–30, both in *Journal of Forestry* 91, no. 12 (1993).

18 William C. McComb, Thomas A. Spies, and William H. Emmingham, "Douglas-Fir Forests: Managing for Timber and Mature Forest Habitat," 31–42, in *Journal of Forestry* 91, no. 12 (1993).

19 Jack Ward Thomas and Susan Huke, "The Forest Service Approach to Healthy Ecosystems," *Journal of Forestry* 94, no. 8 (1996): 14–15.

20 Thomas and Huke, "Forest Service Approach," 16–17.

21 Thomas A. More, "Forestry's 'Fuzzy' Concepts: An Examination of Ecosystem Management," *Journal of Forestry* 94, no. 8 (1996): 19–23.

22 Roger A. Sedjo, "Toward an Operational Approach to Public Forest Management," *Journal of Forestry* 94, no. 8 (1996): 24–27.

23 Sedjo, "Toward an Operational Approach," 27.

24 Chase, *In a Dark Wood*, xiii.

25 Chase, *In a Dark Wood*, xiii–xxii, and 1–2.

26 Chase, *In a Dark Wood*, 7–8.

27 Chase, *In a Dark Wood*, 9–10.

28 Chase, *In a Dark Wood*, 131–135, 148.

29 Max Geier, "Jerry F. Franklin (1936–)," in https://oregonencyclopedia.org/articles/franklin_jerry_f/#.Wyvu9PZFwkE; and Jerry F. Franklin et al., *Ecological Characteristics of Old-Growth Douglas-Fir Forests*, USDA Forest Service, Pacific Northwest Forest and Range Experiment Station, General Technical Report PNW-118 (February 1981).

30 Chase, *In a Dark Wood*, 156–158. For evidence of disturbance research among Andrews scientists, see C. T. Dyrness, *Mass Soil Movements in the H. J. Andrews Experimental Forest*, USDA Forest Service, Pacific Northwest Forest and Range Research Station, Portland, Research Note, PNW-42, 1967; Victoria Ruiz-Villanueva and Markus Stoffel, "Frederick J. Swanson's 1976–1979 Papers on the Effects of Instream Wood on Fluvial Processes and Instream Wood Management," *BioScience* 48, no. 9: 681–689; and James R. Sedell and Clifford N. Dahm, "Catastrophic Disturbances to Stream Ecosystems: Volcanism and Clear-Cut Logging," in *Current Perspectives in Microbial Ecology*, edited by M. J. Klug and C. A. Reddy (Washington, DC: American Society for Microbiology, 1983), 531–539.

31 Chase, *In a Dark Wood*, 372–374.

32 James R. Skillen, *Federal Ecosystem Management: Its Rise, Fall, and Afterlife* (Lawrence: University Press of Kansas, 2015), 17–33.

33 Skillen, *Federal Ecosystem Management*, 388–389.

34 Skillen, *Federal Ecosystem Management*, 411–418.

35 Art McKee to Fred Swanson, June 30, 2018, in William G. Robbins Papers, Special Collections and Archives, Oregon State University (hereafter SCARC).

36 Ted Dyrness et al., "Flood of February 1996 in the H. J. Andrews Experimental Forest," March 27, 1996, H. J. Andrews Bibliography, http://andrewsforest.oregonstate.edu/pubs/webdocs/reports/flood.htm.

37 This information and the quotations are in Jon R. Luoma, *Hidden Forest: The Biography of an Ecosystem* (1999; Corvallis: Oregon State University Press, 2006), 175–179.

38 Sally Duncan, "Lessons from a Flooded Landscape," *Science Findings* (Portland: Pacific Northwest Research Station, February 1998), 1; and *Oregonian*, August 8, 1996.

39 Eugene *Register-Guard*, October 22, 1996.

40 Eugene *Register-Guard*, October 22, 1996.

41 Duncan, "Lessons from a Flooded Landscape," 1–4.

42 Duncan, "Lessons from a Flooded Landscape," 4.

43 F. J. Swanson, J. A. Jones, D. O. Wallin, and J. H. Cissel, "Natural Variability—Implications for Ecosystem Management," in *Ecosystem Management: Principles and Application*, vol. 2, USDA Forest Service, Pacific Northwest Research Station, General Technical Report PNW-GTR-318 (February 1994), 80–82, 90.

44 Carolos Galindo-Leal and Fred L. Bunnell, "Ecosystem Management: Implications and Opportunities of a New Paradigm," *Forestry Chronicle* 7, no. 5 (Sept/Oct 1995): 601–603.

45 Galindo-Leal and Bunnell, "Ecosystem Management," 604–605.

46 Norman L. Christensen et al., "The Report of the Ecological Society of America Committee on the Scientific Basis for Ecosystem Management," *Ecological Applications* 6, no. 3 (1996): 667–676.

47 Jack Ward Thomas, "Foreword," in *Ecosystem Management*, edited by Alan Haney and Mark Boyce (New Haven, CT: Yale University Press, 1997), ix–xl.

48 Jerry F. Franklin, "Ecosystem Management: An Overview," in *Ecosystem Management*, 21–23.

49 Franklin, "Ecosystem Management," in *Ecosystem Management*, 24–30.

50 Skillen, *Federal Ecosystem Management*, 87–114.

51 Gene E. Likens, "Limitations to Intellectual Progress in Ecosystem Science," in *Successes, Limitations and Frontiers in Ecosystem Science*, edited by P. M. Groffman and M. L. Place (Springer: New York, 1998), 241, 248–249, 254.

52 Likens, "Limitations to Intellectual Progress," 256–266. For provocative insights to the intersection between science and policy, see Harry N. Scheiber, "From Science to Law to Politics: An Historical View of the Ecosystem Idea and Its Effect on Resource Management," *Ecology LQ* 24 (1997): 631–651. The author says that there is a strong argument for scientists to avoid "political combat over how to apply their findings to actual management practices" (644)

53 Richard Haeuber, "Setting the Environmental Policy Agenda: The Case of Ecosystem Management," *Natural Resources Journal* 36, no. 1 (Winter 1996): 10–19.

54 David N. Bengston, George Xu, and David P. Fan, "Attitudes toward Ecosystem Management in the United States, 1992–1998," *Society and Natural Resources* 14 (2001): 471–473, 481–482.

55 Robert V. O'Neill, "Is It Time to Bury the Ecosystem Concept? (With Full Military Honors of Course!)," *Ecology* 82, no. 12 (2001): 3275–3276.

56 O'Neill, "Is It Time?" 3276–3277; and Robert V. O'Neill and James R. Kahn, "*Homo economus* as a Keystone Species," *BioScience* 50, no. 4 (April 2000): 333–337.

57 O'Neill, "Is It Time?" 3279–3282.

58 George Hoberg, "Science, Politics, and US Forest Service Law: The Battle over the Forest Service Planning Rule," *Natural Resource Journal* 44 (Winter 2004): 11–15.

59 K. Norman Johnson et al., "Sustaining the People's Lands: Recommendations for Stewardship of the National Forests and Grasslands into the Next Century," *Journal of Forestry* 97, no. 5 (1999): 6–12.

60 Johnson et al., "Sustaining the People's Lands," 6–9.

61 Hoberg, "Science, Politics, and US Forest Service Law," 16–25.

62 Tomas M. Koontz and Jennifer Bodine, "Implementing Ecosystem Management in Public Agencies: Lessons from the US Bureau of Land Management and the Forest Service," *Conservation Biology* 22, no. 1 (February 2008): 63–66.

63 Skillen, *Federal Ecosystem Management*, 261–263.

64 Skillen, *Federal Ecosystem Management*, 263–264.

65 Skillen, *Federal Ecosystem Management*, 266–270.

CHAPTER 6

1 Richard Hodson, "Climate Change," *Nature* 550, no. 7675 (October 2017): S62.

2 H. J. Andrews Forest LTER—2000 Annual Report—LTER, Year 4, pp. 1–5, in William G. Robbins Papers, Special Collections and Archives, Oregon State University (hereafter SCARC).

3 Andrews LTER5 Proposal, Submitted to the National Science Foundation (February 2002), 83, in SCARC.

4 Andrews LTER5 Proposal (February 2002), 84.

5 H. J. Andrews Forest LTER—2000 Annual Report—LTER4, Year 4; and H. J. Andrews Forest LTER—2001 Annual Report—LTER4, Year 5, both in SCARC.

6 H. J. Andrews Forest LTER—2000 Annual Report—LTER4, Year 4, in SCARC.

7 H. J. Andrews Forest LTER—2001 Annual Report—LTER4, Year 5, in SCARC.

8 H. J. Andrews Forest LTER—2000 Annual Report—LTER 4, Year 4; and H. J. Andrews Forest LTER—2001 Annual Report—LTER4, Year 5, in SCARC.

9 Denise Lach, Peter List, Brent Steel, and Bruce Shindler, "Advocacy and Credibility of Ecological Scientists in Decision Making: A Regional Study," *BioScience* 53, no. 2 (February 2003): 170–172.

10 *International Long-Term Ecological Research Network* (compiled by the US LTER Office, Albuquerque, NM, 1998), 74–81, in SCARC.

11 Swanson_LTERhistory_communities_081018, Fred Swanson book notes, 1–2, and 6–11, in SCARC.

12 "Project Summary," Andrews LTER5 Proposal (submitted to NSF February 2002), 1–4, in SCARC.

13 "Project Summary" (February 2002), 33.

14 Long-Term Ecological Research at the H. J. Andrews Experimental Forest (LTER5), Annual Report for Period 11/2008–10/2009, Barbara Bond, principal investigator (Award ID 0218088), 1–3, in SCARC.

15 Long-Term Ecological Research, Annual Report for Period 11/2008–10/2009, 9–10, in SCARC.

16 Long-Term Ecological Research, Annual Report for Period 11/2008–10/2009, 1–3, and 9–10.

17 Andrews Forest LTER6: Proposal to the National Science Foundation, Barbara Bond, principal investigator, February 2008, Cover Sheet and Project Summary, in SCARC.

18 Andrews Forest LTER6: Proposal, February 2008, Cover Sheet and Project Summary, 22–24.

19 Andrews Forest LTER6: Proposal, February 2008, Cover Sheet and Project Summary, 25 and 32.

20 The records of the symposia can be found at https://andrewsforest.oregonstate.edu/outreach/events/symposium.

21 Eugene Natural History Society 1998 Lecture Series; and Free Public Tour Offered—Saturday, September 12, both in SCARC.

22 Corvallis *Gazette-Times*, August 24, 1998; Planned and Proposed Andrews Forest 50th Anniversary Activities for August 21, 1998; H. J. Andrews Experimental Forest 50th Anniversary Celebration, August 21, 1998, Andrews Forest Headquarters; and USDA Forest Service, Pacific Northwest Research Station, *Research News* (Oct/Nov 1998), all in SCARC.

23 H. J. Andrews Experimental Forest 50th Anniversary Celebration, in SCARC.

24 "LTER6: Using Innovative Approaches and Long-Term Research to Address Complex Socio-Ecological Questions, 10th Annual Symposium, H. J. Andrews Experimental Forest," April 15, 2009, in SCARC.

25 "Networks and Synthesis, H. J. Andrews Experimental Forest, Long-Term Ecological Research (LTER), 11th Annual Symposium," April 20, 2010, in SCARC.

26 "H. J. Andrews Experimental Forest, Long-Term Ecological Research Program (LTER) Symposium, Long-Term Research into the Future," January 29, 2015; and "H. J. Andrews Experimental Forest, Long-Term Ecological Research (LTER) 2016 Symposium," May 13, 2016, both in SCARC.

27 Fred Swanson to the author, January 3, 2018, in SCARC.

28 Hannah Gosnell, Michael Paul Nelson, Julia Jones, and Fred Swanson, Divergent Uses of Long-Term Ecological Research in Debates over Forest Values and Forest Governance, rough draft of a paper courtesy of Fred Swanson, no pagination; and Andrew Gray to the author, November 15, 2018, both in SCARC.

29 Gosnell et al., Divergent Uses, no pagination, in SCARC.

30 G. M. Woodwell et al., "Global Deforestation: Contribution to Atmospheric Carbon Dioxide," *Science* 222, no. 4628 (1983): 1081–1086; James Hansen and Sergei Lebedeff, "Global Trends of Measured Surface Air Temperature," *Journal of Geophysical Research* 92, no. D11 (1987): 13345–13372; Richard A. Houghton and George M. Woodwell, "Global Climate Change," *Scientific American* 260, no. 4 (April 1989): 36–47; and Stephen A. Schneider, "The Changing Climate," *Scientific American* 260, no 3 (September 1989): 70. For a brief account of Hansen's career, see Seth Borenstein, "James Hansen Wishes He Wasn't So Right about Global Warming" (June 2018), https://phys.org/news/2018-06-james-hansen-wasnt-global.html.

31 David A. Perry and Jeffrey G. Borchers, "Climate Change and Ecosystem Responses," *Northwest Environmental Journal* 6 (1990): 293–313; F. J. Swanson et al., "Some Emerging Issues in Watershed Management: Landscape Patterns, Species Conservation, and Climate Change," in *Watershed Management* (New York: Springer, 1992), 307–323; and George A. King and David T. Tingey, *Potential Impacts of Climate Change on Pacific Northwest Vegetation* Corvallis, Oregon, US Department of Energy, Office of Scientific and Technical Information, https://www.osti.gov/biblio/5178953, p. iii.

32 David Greenland, "Regional Context of the Climate of the H. J. Andrews Experimental Forest, Oregon," *Proceedings of the Tenth Annual Pacific Climate Workshop*, Interagency Ecological Studies, Technical Report 36 (California Department of Water Resources, 1993), 41–42, 54.

33 The Spring Creek Project for Ideas, Nature, and the Written Word originated at Oregon State University in 2003, with cofounder Kathleen Dean Moore serving as director for the first ten years. Moore, a Distinguished Professor of Philosophy, left the university in 2013 to pursue professional interests in environmental issues.

34 Mathew C. Nesbit, Mark A. Hixon, Kathleen Dean Moore, and Michael Nelson, "Four Cultures: New Synergies for Engaging Society on Climate Change," *Frontiers in Ecology and Environment* 8 (2001): 329–331.

35 Oregon Climate Change Research Institute, *Oregon Climate Assessment Report: Legislative Summary* (Salem, December 2010), 1–2.

36 Thomas A. Spies et al., "Climate Change Adaptation Strategies for Federal Forests on the Pacific Northwest, USA: Ecological, Policy, and Socio-Economic Perspectives," *Landscape Ecology* 25 (2010): 1185–1186.

37 Spies et al., "Climate Change Adaptation Strategies," 1186–1187.

38 Spies et al., "Climate Change Adaptation Strategies," 1195–1197.

39 David R. Foster, "Expanding the Integration and Application of Long-Term Ecological Research," 323; and Alan K. Knapp et al., "Past, Present, and Future Roles of Long-Term Experiments in the LTER Network," 377–389, both in *BioScience* 62, no. 4 (special issue, April 2012).

40 Charles T. Driscoll et al., "Science and Society: The Role of Long-Term Studies in Environmental Stewardship," 364; and G. Philip Robertson et al., "Long-Term Ecological Research in a Human-Dominated World," 351, both in *BioScience* 62, no. 4 (special issue, April 2012).

41 Jonathan Thompson, "Scenario Studies as a Synthetic Activity for Long-Term Ecological Research," *BioScience* 62, no. 4 (special issue, April 2012): 367–376.

42 Heejun Chang and Julia Jones et al., "Climate Change and Freshwater Resources in Oregon," in *Oregon Climate Assessment Report*, College of Oceanic and Atmospheric Sciences (Oregon State University, 2010), 71–151.

43 Julia A. Jones, "Hydrologic Responses to Climate Change: Considering Geographic Context and Alternative Hypotheses," *Hydrological Processes* 25, no. 12 (2011): 1996–2000; and Kendra L. Hatcher and Julia A. Jones, "Climate and Streamflow Trends in the Columbia River Basin: Evidence for Ecological and Engineering Resilience to Climate Change," *Atmosphere-Ocean* 51, no. 4 (2013): 436–455.

44 Kathleen M. Moore, "Optimizing Reservoir Operations to Adapt to 21st Century Expectations of Climate and Social Change in the Willamette Basin, Oregon" (PhD diss. in geography, Oregon State University, 2015), abstract, and 180–182.

45 Eric A. Sproles, Travis R. Roth, and Anne Nolin, "Future Snow? A Spatial-Probabilistic Assessment of the Extraordinarily Low Snowpacks of 2014 and 2015 in the Oregon Cascades," *The Cryosphere* 11, no. 1 (2017): 331–332.

46 Sproles et al., "Future Snow?" 332.

47 Sproles et al., "Future Snow?" 332–333.

48 Anna Young, "The Effect of Climate Change on Pollinators and the Implications for Global Agriculture: A Case Study in the H. J. Andrews Experimental Forest, Oregon" (Yale College, Senior Essay in Environmental Studies, 2016), abstract, and 56–59, in SCARC.

49 Chelsea Batavia and Michael Paul Nelson, "Translating Climate Change Policy into Forest Management Practice in a Multiple-Use Context: The Role of Ethics," *Climatic Change* 148, no. 1–2 (2018): 81–94, published online March 28, 2018, in SCARC.

50 Jerry F. Franklin and K. Norman Johnson, "A Restoration Framework for Federal Forests in the Pacific Northwest," *Journal of Forestry* 110, no. 8 (2012): 429.

51 Franklin and Johnson, "A Restoration Framework," 429–437.

52 Katerina Honzakova, "Small-Scale Variations of Climate Change in Mountainous-Forested Terrain," master's thesis, Global Change and Ecology (University of Bayreuth, 2017), I, 1–2, 7, 35–36; and Mark Schulze to the author, January 7, 2019, both in SCARC. The author assumed that Honzakova had spent time at the Andrews.

53 David M. Bell, Thomas A. Spies, and Robert Pabst, "Historical Harvests Reduce Neighboring Old-Growth Basel Area across a Forest Landscape," *Ecological Applications* 27, no. 5 (2017): 1666–1667.

54 Bell et al., "Historical Harvests," 1167–1168, 1671–1674.

55 Mathew J. Kaylor and Dana R. Warren, "Canopy Closure after Four Decades of Postlogging Riparian Forest Regeneration Reduces Cutthroat Trout Biomass in Headwater Streams through Bottom-Up Pathways," *Canadian Journal of Fisheries and Aquatic Sciences* 75, no. 4 (2017): 513–524.

56 Thomas A. Spies et al., *Synthesis of Science to Inform Land Management within the National Forest Plan Area: Executive Summary*, USDA Forest Service, Pacific Northwest Research Station, Portland, Oregon, General Technical Report, PNW-GTR-970 (August 2018), abstract.

57 Spies et al., *Synthesis of Science*, 8–9; and Jack Ward Thomas et al., "The Northwest Forest Plan: Origins, Components, Implementation Experience, and Suggestions for Change," *Conservation Biology* 20, no. 2 (2006): 281.

58 Spies et al., *Synthesis of Science*, 13–19.

59 Spies et al., *Synthesis of Science*, 184.

60   Spies et al., *Synthesis of Science*, 184–185.

CHAPTER 7

1   *Kirkus Reviews*, February 1, 2018, unsigned review of *The Overstory*, by Richard Powers, https://www.kirkusreviews.com/book-reviews/richard-powers/the-overstory/.

2   William G. Robbins, *Colony and Empire: The Capitalist Transformation of the American West* (Lawrence: University Press of Kansas, 1994), 66–68; Frederick J. Swanson, "Confluence of Arts, Humanities, and Science at Sites of Long-Term Ecological Inquiry," *Ecosphere* 6, no. 8 (2016): 2–4; and Charles Goodrich and Frederick J. Swanson, "Long-Term Ecological Reflections: Art among Science among Place," *Terrain.org: A Journal of the Built and Natural Environments* (2016), https://www.terrain.org/2016/guest-editorial/long-term-ecological-reflections/. For an important new book on field research in the United States, see Michael J. Lannoo, *This Land Is Your Land: The Story of Field Biology in America* (Chicago: University of Chicago Press, 2018).

3   Robert Michael Pyle, *Wintergreen: Rambles in a Ravaged Land* (New York: Scribner's, 1986); and "Interview with Robert Michael Pyle," *Terrain.org: A Journal of the Built and Natural Environment*, https://www.terrain.org/2015/interviews/robert-michael-pyle/.

4   William Dietrich, *The Final Forest: The Battle for the Last Great Trees of the Pacific Northwest* (New York: Simon and Schuster, 1992); E. Maass, *Library Journal* 117, no. 8 (May 1, 1992): 108; and *Kirkus Reviews*, April 15, 1992, unsigned review of *The Final Forest: The Battle for the Last Great Trees of the Pacific Northwest*, by William Dietrich, https://www.kirkusreviews.com/book-reviews/william-dietrich/the-final-forest/.

5   Gerald Stern's praise for the Whitman Award and the LSU Press citation can be found on its website, https://alisonhawthornedeming.com/books/poetry/science-and-other-poems/.

6   Swanson, "Confluence of Arts," 17–18; Andrew C. Gottlieb, "Ecological Reflections: About Earth Scientist Fred Swanson," *Terrain.org: A Journal of the Built and Natural Environments* (November 10, 2013), https://www.terrain.org, 8; and "The Spring Creek Project for Ideas, Nature, and the Written Word," College of Liberal Arts, Oregon State University, https://liberalarts.oregonstate.edu/centers-and-initiatives/spring-creek-project/.

7   Gottlieb, "Ecological Reflections: About Earth Scientist Fred Swanson"; and Lissy Goralnik et al., "Arts and Humanities Efforts in the US Long-Term Ecological Research (LTER) Network: Understanding Perceived Values and Challenges," in *Earth Stewardship: Linking Ecology and Ethics in Theory and Practice*, edited by Ricardo Rozzi et al. (New York: Springer 2015), 250, https://www.springer.com/us/book/9783319121321.

8   Swanson, "Confluence of Arts," 1, 5–8.

9   David R. Foster, ed., *Hemlock: A Forest Giant on the Edge* (New Haven, CT: Yale University Press, 2014), xiv–xv, 71.

10   "Long-Term Ecological Reflections," *Network News* (April 2005), LTER Intranet, http://lternet,edu/news/Article4.html.

11   Frederick J. Swanson, Charles Goodrich, and Kathleen Dear Moore, "Bridging Boundaries: Scientists, Creative Writers, and the Long View of the Forest," *Frontiers in Ecology and the Environment* 6, no. 9 (2008): 499–504.

12   Jonathan Thompson, "Long-Term Ecological Reflections: Writers, Philosophers, and Scientists Meet in the Forest," *Science Findings*, no. 105 (August 2008), Portland, Pacific Northwest Research Station, 1–5.

13   Swanson, "Confluence of Arts," 15–20.

14   Robin Kimmerer, biography and links to writings, Andrews website, Andrews Forest Log, https://liberalarts.oregonstate.edu/centers-and-initiatives/spring-creek-project/programs-and-residencies/long-term-ecological-reflections/forest-log.

15   Kimmerer, "Interview with a Watershed," Andrews Forest Log.

16   Deming, "Attending to the Beautiful Mess of the World," in *The Way of Natural History*, edited by Thomas Lowe Fleischner (San Antonio, TX: Trinity University Press, 2011), 174–179.

17   Deming, "Attending to the Beautiful Mess," 179–181.

18   Kathie Durbin, "Reflections on Change, Natural, and Otherwise: A Forest Journal," Lookout Creek, May 11, 2007, Andrews Forest Log.

19   Durbin, "Reflections on Change."

20   Bill Sherwonit, "Reflections on Thrush Songs, Newt Tracks and Old-Growth Stands of Trees," *Interdisciplinary Studies in Literature and Environment* 16, no. 4 (Autumn 2009): 823–827.

21   Sherwonit, "Reflections," 830–832.

22 Tim Fox, "Primordial Chords," Andrews Forest Log.

23 Tim Fox, "Barred Owls and Belonging," Andrews Forest Log.

24 Swanson et al., "Bridging Boundaries," 499–504; Jeff Baker, "Reflection, Renewal in 'Blast Zone,'" May 16, 2008, jbaker@news.oregonian.com; and Charles Goodrich, Kathleen Dean Moore, and Frederick J. Swanson, eds., *In the Blast Zone: Catastrophe and Renewal on Mount St. Helens* (Corvallis, OR: Oregon State University Press, 2008).

25 Ursula K. Le Guin, "Coming Back to the Lady," in Goodrich et al., *In the Blast Zone*, 2–4.

26 Le Guin, "Coming Back," in Goodrich et al., *In the Blast Zone*, 2–8.

27 Kathleen Dean Moore, "In Endless Song," in Goodrich et al., *In the Blast Zone*, 22–25.

28 Moore, "In Endless Song," in Goodrich et al., *In the Blast Zone*, 26–27.

29 Charles Goodrich, "A Gardener Goes to the Volcano," in Goodrich et al., *In the Blast Zone*, 56–61.

30 Christine Colasurdo, "Everlasting Wilderness," in Goodrich et al., *In the Blast Zone*, 77–83.

31 Colasurdo, "Everlasting Wilderness," in Goodrich et al., *In the Blast Zone*. The poems are on pages 17–20, 36–40, 52–53, 104, 114–115.

32 Judith L. Li and M. L. Herring, *Ellie's Log: Exploring the Forest Where the Tree Fell* (Corvallis: Oregon State University Press, 2013); Susan Palmer, "Children's Book Set in Familiar National Forest," Eugene *Register-Guard*, April 2, 2013; and Oregon State University Press, press release, March 25, 2013, "OSU Press Publishes Its First Book for Children," https://today.oregonstate.edu/archives/2013/mar/osu-press-publishes-its-first-book-children.

33 NSF LTER Network, Schoolyard Book Series, https://lternet.edu; and Judith L. Li and M. L. Herring, *Ricky's Atlas: Mapping a Land on Fire* (Corvallis: Oregon State University Press, 2016).

34 Nathaniel Brody, Charles Goodrich, and Frederick J. Swanson, eds., *Forest Under Story: Creative Inquiry in an Old-Growth Forest* (Corvallis: Oregon State University, 2016); Goodrich, "Entries into the Forest," 5–14, in *Forest Under Story*.

35 The Groundwork essays are on the following pages of *Forest Under Story* (2016): 35, 84, 120, 135, 152, 197, 212.

36 Freeman House, "A Watershed Runs Through You," *Yes! Journalism for People Building a Better World*, December 31, 2003, https://www.yesmagazine.org/issues/whose-water/a-watershed-runs-through-you; and Freeman House, "Varieties of Attentiveness," in *Forest Under Story*, 93–96.

37 Laird Christensen, "The Other Side of the Clear-Cut," in *Forest Under Story*, 137–139.

38 Christensen, "The Other Side of the Clear-Cut," in *Forest Under Story*, 139–141; Green Mountain College closed in 2019.

39 Christensen, "The Other Side of the Clear-Cut," in *Forest Under Story*, 142–147.

40 Frederick H. Swanson, "Field Notes I," Andrews Forest Log, 1–2.

41 Swanson, "Field Notes I," 7–8.

42 Swanson, "Field Notes I," 10.

43 Aaron Ellison, "Decomposition and Memory," Andrews Forest Log.

44 Ellison, "Decomposition and Memory."

45 Vincent J. Miller, "*Laudato Si'* in an Old Growth Forest," Andrews Forest Log.

46 Miller, "*Laudato Si.*"

47 Ann Rosenthal, "Artist Residency 2018," Andrews Forest Log.

48 Rosenthal, "Artist Residency 2018."

49 Lissy Goralnik et al., "Arts and Humanities Efforts in the US Long-Term Ecological Research (LTER) Network," in *Earth Stewardship: Linking Ecology and Ethics in Theory and Practice*, edited by Ricardo Rozzi et al. (New York: Springer, 2015), 250–253. doi:10.1007/978-3-319-12133-8_16.

50 Goralnik, "Arts and Humanities Efforts," 250–253.

51 Lissy Goralnik, Michael Paul Nelson, Hannah Gosnell, and Mary Beth Leigh, "Arts and Humanities Inquiry in the Long-term Ecological Research Network: Empathy, Relationships, and Interdisciplinary Collaborations," *Journal of Environmental Studies* 7, no. 2 (2017): 361–373.

52 Fred Swanson to Nick Houtman, OSU Director of Communications, September 5, 2005; and Swanson to the author, February 17, 2019, both in SCARC.

53 "NSF Director Praises LTER," LTER, *The Network News* (April 2005), http://www.lternet.edu/news/PrintArticle5.html.

54 Stephen T. Jackson, "Humboldt for the Anthropocene," *Science* 365 (September 13, 2019), 1074–1075; and https://andrewsforest.oregonstate.edu/outreach/arts-and-humanities.

CONCLUSION

1   Edward O. Wilson, *Half-Earth: Our Planet's Fight for Life* (New York: Liveright Publishing, 2016), 15.

2   Max Geier, interview with Roy Silen, September 9, 1996; Geier, *Necessary Work: Discovering Old Forests, New Outlooks, and Community on the H. J. Andrews Experimental Forest, 1948–2000*, USDA Forest Service, Pacific Northwest Research Station, General Technical Report, PNW-GTR-687 (2007), 16; The Obsidians, Eugene, Oregon, http://www.obsidians.org/; Constance J. Burke, Historic Fires in the Central Western Cascades, Oregon, master's thesis, Oregon State University, 1979, 6–79; and Fred Swanson to the author, February 27, 2019, in SCARC.

3   Warren Cornwall, "Against the Grain," *Science* 358, no. 6359 (October 2017): 24; and Mark E. Swanson et al., "The Forgotten State of Forest Succession: Early-Successional Ecosystems on Forest Sites," *Frontiers in Ecology and the Environment* 9, no. 2 (2011): np, www.frontiersinecology.org.

4   Swanson et al., "Forgotten Stage," 117–125; and Cornwall, "Against the Grain," 26.

5   Cornwall, "Against the Grain," 26–27.

6   Jerry F. Franklin, K. Norman Johnson, and Debora L. Johnson, *Ecological Forest Management* (Long Grove, IL: Waveland Press, 2018), https://www.waveland.com/browse.php?t=730&r=d|103; and Bert Loosmore, "Review of *Ecological Forest Management* by Jerry Franklin," in *Conservation Northwest*, November 2018, https://www.conservationnw.org/ecological-forest-management-review/.

7   V. Alaric Sample, review of *Ecological Forest Management*, in *Journal of Forestry* 116, no. 5: 487–488; and *Ecological Forest Management*, reviewed in Andy Kerr's *Public Lands Blog*, http://www.andykerr.net/kerr-public-lands-blog/.

8   Frederick J. Swanson and Charles M. Crisafulli, "Volcano Ecology: State of the Field and Contributions of Mount St. Helens Research," in *Ecological Responses at Mount St. Helens: Revisited 35 Years after the 1980 Eruption*, edited by Virginian H. Dale and Charles Crisafulli (New York: Springer, 2017), 317–318. For the twenty-five-year assessment of St. Helens research, see Virginian H. Dale, Frederick J. Swanson, and Charles Crisafulli, eds., *Ecological Responses to the 1980 Eruption of Mount St. Helens* (New York: Springer, 2005).

9   Wilson, *Half-Earth*, 174.

10   The author's conversation with Mark Harmon, March 25, 2019.

11   The author's conversation with Tom Spies, April 4, 2019.

12   The author's conversation with Barbara Bond, April 16, 2019.

13   The author's conversations with Mark Schulze, May 4, 2019.

14   The author's conversation with Fred Swanson, May 15, 2019.

15   Fred Swanson, conversations with the author, April 29, 2019.

16   Barbara Kingsolver, Review of *The Overstory* in the *New York Times*, April 9, 2019.

17   Kathleen Dean Moore, *Overstory*, a Random Review at the Corvallis-Benton County Library, January 12, 2019.

18   Jake Rice et al., eds., "Foreword," in *The IPBES Regional Assessment Report on Biodiversity and Ecosystem Services for the Americas* (Bonn: United Nations, Secretariat of the Intergovernmental Science-Policy Platform on Biodiversity and Ecosystem Services, 2019), iv–v.

Index